Max Marckwald, Thomas A. Haig

The Movements of Respiration and Their Innervation in the Rabbit

with a supplement on the relation of respiration to deglutition, and on the question

of the existence of respiratory centres in the spinal cord

Max Marckwald, Thomas A. Haig

The Movements of Respiration and Their Innervation in the Rabbit
with a supplement on the relation of respiration to deglutition, and on the question of the existence of respiratory centres in the spinal cord

ISBN/EAN: 9783337198411

Printed in Europe, USA, Canada, Australia, Japan

Cover: Foto ©berggeist007 / pixelio.de

More available books at **www.hansebooks.com**

THE MOVEMENTS OF RESPIRATION.

THE

MOVEMENTS OF RESPIRATION

AND

THEIR INNERVATION IN THE RABBIT.

WITH A SUPPLEMENT
ON THE RELATION OF RESPIRATION TO DEGLUTITION,
AND ON THE QUESTION OF THE EXISTENCE OF RESPIRATORY CENTRES IN
THE SPINAL CORD.

BY

MAX MARCKWALD, M.D.

PHYSICIAN AT THE BATHS OF KREUZNACH.

TRANSLATED BY
THOMAS ARTHUR HAIG, STUDENT OF MEDICINE, UNIVERSITY OF GLASGOW;

AND REVISED BY THE AUTHOR.

WITH

AN INTRODUCTORY NOTE

BY JOHN G. M‘KENDRICK, M.D., LL.D., F.R.S.,
PROFESSOR OF PHYSIOLOGY, UNIVERSITY OF GLASGOW.

LONDON:
BLACKIE & SON, 49 & 50 OLD BAILEY, E.C.
GLASGOW, EDINBURGH, AND DUBLIN.
1888.

RESEARCHES FROM THE PHYSIOLOGICAL LABORATORY
OF THE UNIVERSITY OF BERNE.

OTHER WORKS BY DR. MARCKWALD:—

1. Ueber Verdauung und Resorption im Dickdarme des Menschen.

2. Ueber die Kegelmantelförmige Excision der Vaginalportion u. ihre Anwendung.

3. Ueber die Wirkungen von Ergotin, Ergotinin u. Sclerotinsäure auf Blutdruck, Uterusbewegungen u. Blutungen.

INTRODUCTORY NOTE.

The innervation of the respiratory movements is well known to be one of great complexity, and there are many vague and unsatisfactory statements regarding it in current literature. More especially is this the case with reference to the questions as to whether there are respiratory centres in the spinal cord as well as in the *medulla oblongata*, and as to how far and in what manner these centres are influenced by nervous impressions flowing from the periphery of the body, or passing from the higher nervous centres downwards.

In 1811, Legallois first demonstrated that an important centre for respiratory movements existed in the *medulla oblongata*, and Flourens, with whose name this discovery is usually associated, did little except to give to this centre the well-known name, *nœud vital*. Since then, physiological experiment has been chiefly occupied in limiting and defining this centre. Thus Longet, Charles Bell, Schiff, Giercke, and others, endeavoured to define with accuracy certain centres or bundles of nerve fibres as specially constituting the respiratory centre.

The next step was made in 1860, when Brown-Séquard stated that respiratory movements might take place after removal of the *medulla oblongata*, and this statement was supported by Dr. Bennett-Dowler of New Orleans, who noted the fact in decapitated crocodiles, by Dr. B. W. Richardson, who made the observation on new-born mammals, and by Rokitansky, who, in 1876, noticed respiratory movements in decapitated young rabbits after the injection of strychnia. This led to a remarkable series of experiments by Langendorff, dating from 1880 to the present year, in which he accumulates facts in favour of the view that respiratory centres exist in the spinal cord, and that these can become active in a reflex or automatic manner. This, then, is one aspect of the problem of breathing.

From the earliest times, the influence of sudden and strong irritation of the skin must have been observed to have an influence on
respiratory movements, but it was not until 1822 that Sir Charles
Bell, in his renowned investigations, showed that these irritations acted
through the respiratory centre.   Dr. Marshall Hall, in his doctrine of
reflex action, gave clearness and precision to this statement.   Years
passed, and in 1855 Brown-Séquard attributed the sudden death
following section of the *medulla oblongata* to an inhibitory action
which arrested both cardiac and respiratory movements.   An arrest
or inhibition of the respiratory centre was observed by Goltz in 1867
to be also caused by strong stimulation of the abdominal nerves.   It is
well known that other reflexes may be checked in a similar way by
peripheral impressions.   These observations were the beacon-lights that
guided many investigators in the search for the paths by which peripheral impressions pass to the respiratory centres, and led to various
well-known theories of respiratory action.   Such speculative views,
founded on experimental data, culminated in the well-known theory of
Rosenthal, announced in 1861, in which he assumes the existence both
of inspiratory and of expiratory centres in the medulla, the expiratory
being stimulated by impressions transmitted largely through the superior laryngeal branch of the vagus, whilst the inspiratory is affected
by impressions coming from the lungs by the pulmonary branches of
the vagus.   Further, it was shown that both centres may be influenced
by the nature of the blood passing through them, and by impressions
conducted by the sensory nerves of the skin, and even by those of
special sense.   How far these views have been justified is a second
aspect of the problem of breathing.

    Lastly, there is the profound question as to what is the cause of
rhythm.   The rhythmic movements of cilia, the rhythmic movements
of the bodies of many invertebrates, the rhythmic movements of certain
muscles without the mechanism of nerve centres, and the rhythmic
movements of the heart, probably involve the same kind of molecular
mechanism as determines the rhythmic discharges of the respiratory
centres.   What is it that determines the sudden outburst of nervous

currents that call into action the muscles of inspiration? Is this due
to a (chemical) rearrangement of the molecules of the nerve centre by
which a state of tension is reached, when the energy becomes kinetic, in
the form of a discharge, or does there occur in the centre a storing up
or summation of peripheral impressions by which energy is, as it
were, gradually accumulated, until it finds vent in motor impulses
streaming out to the muscles of inspiration? How are these obscure
processes influenced by impressions coming from the periphery? Does
rhythm depend on central changes alone, or can peripheral impressions
in some way originate intermittent action? This is a third view of the
problem of breathing.

All these questions are critically discussed in the able monograph
by Dr. Marckwald now laid before English readers. After numerous
and carefully conducted experiments, Dr. Marckwald has arrived at
new theories as to the action of the respiratory centres, the rhythm of
respiration caused by the vagi, and the action of the glossopharyngei,
and other nerves on the respiratory centre. A perusal of the original
paper led me to form the opinion that it ought to be known in this
country, more especially by physicians, to whom any advance in the
knowledge of nervous mechanisms must be invaluable, and, with Dr.
Marckwald's consent, I asked Mr. Thomas Arthur Haig, one of my
students, to prepare the present translation. This he has done, perhaps,
in too literal a form, but the literal rendering has conserved some of
the spirit of Dr. Marckwald's original paper. The translation has been
revised by Dr. Marckwald, who has added largely to the valuable
bibliography at the end. Two appendices have also been added. The
first appendix relates to the interesting relations that have been found
to exist between the mechanism of respiration and the mechanism of
deglutition, and the other contains a critical examination by Dr.
Marckwald of the much-debated question as to whether there are or
are not respiratory centres in the spinal cord.

The paper is an excellent example of the true experimental method,
guided by sound knowledge of the work of others, and checked by a
severe criticism of results. It is right to mention that in this investi-

gation none of the animals suffered pain. Dr. Marckwald informs me
that, in all the experiments on deglutition, the animal was under the
influence of large doses of morphia from the beginning of the experiment until its death, and that, in the experiments on respiration, the
animal was first put under the influence of morphia, and then the
medulla oblongata was divided, so that all consciousness of peripheral
impressions was abolished. Dr. Marckwald's research is a specimen of
a physiological inquiry far too uncommon, and now almost impossible,
in this country, owing to legislative restrictions. Fortunately, however, for humanity, the results of scientific inquiry are not limited to
the land in which they were first attained.

UNIVERSITY OF GLASGOW,
June, 1888.

CONTENTS.

LIST OF ILLUSTRATIONS.

INNERVATION OF RESPIRATION.

INTRODUCTION.

SINCE the days of Galen, who was the first to describe it more minutely, the theory of respiration shows a literature such as would fill many large volumes. It is to be regretted that the explanation has not kept pace with the number of treatises, and to-day we are still as far from the final answer to the question, What is the cause of respiration? as were Legallois and Flourens, who first pointed out the position of the centre of respiration, and as Marshall Hall and A. W. Volkmann, who by means of numerous and remarkable experiments founded their theories. In fact, one can only say that, on the whole, since time and ability have been spent on the solution of this problem, valuable facts have been won for science in the department of respiration, but contradictions have also greatly increased in number. Even that which a few years ago was regarded as the firm and unassailable property of science, which had been proven by numerous physiological experiments and appeared to be supported by observations on the living subject—namely, the knowledge of the position and simplicity of the centre of respiration,—is to-day forced to yield to other opinions. In place of the venerable " nœud vital " we have whole rows of new centres which are localized in the cerebro-spinal axis from the third ventricle of the brain down into the spinal cord as far as the origin of the phrenic nerves; indeed, the effort to decentralize has gone so far that different groups of respiratory muscles have been allotted "centres" of their own whence their innervation is regulated. Just as little agreement is shown in opinions regarding the excitability possessed by these centres (their so-called automatism) and their natural stimuli (through the blood or through nervous influences, &c.), as well as regards the part played, in the accomplish-

A

ment of respiration, by the individual centripetal nerve tracts in connection with their centres. The subject of the innervation of respiration, therefore, appears darker and more complicated than ever, and when I now make public my work " On Respiration and its Innervation," which has occupied me for many years and is founded on numerous observations, I do so in the hope that through it a greater unison may be again brought about in opinions, and especially I desire to put an end to all the extravagant views which nowadays are advanced with regard to centres and central function. In doing this, I shall as far as possible avoid all theories, and only draw conclusions where they are clearly proved by experiments. To critically examine and work out the scattered literary material would be a very difficult task, and one which I will not attempt. The bibliography at the end of this treatise is by no means complete. I have endeavoured, as far as I could, to take into consideration chiefly those papers which treat of the innervation of the diaphragm, and this I have done in the alphabetical order of the authors' names. In this way it will be easy for the reader, without special mention in the text, to find, with the help of the bibliography, the desired proofs of any statement made by another author and quoted by me.

The first results of investigations by Prof. H. Kronecker of Berne and myself have been already briefly communicated to the Berlin Physiological Society, in the years 1879 and 1880, and may be found in the *Transactions* of that Society. The extended work, which I now make public, and which partly contains the improvement and completion of investigations carried on at the above time, has been made more complete by many new experiments, which I undertook during the winter of 1885–6, in the Physiological Institute at Berne, and in which Professor H. Kronecker has again constantly aided me with his advice.

As animals suitable for experiment, rabbits were chosen, for the reason that they, in the normal as well as under altered conditions (except when they are affected by extreme dyspnœa), breathe exclusively with the diaphragm. I employed a very simple method to register with exactitude the movements of the diaphragm, and thus obtained tracings which, although they did not express the results in absolute measure, gave results such as could be compared both with regard to capacity of respiration and to the muscular effort of respiration. Complicated apparatus was thus unnecessary. Only in a few exceptions did I use dogs, cats, and marmots, when the arrangement of the experiment was altered as far as necessary.

In the following pages there will be considered in their order:

1. The diaphragm and the phrenic nerves.
2. The respiration tracts in the spinal cord.
3. The centres of respiration in the *medulla oblongata*.
4. The centripetal paths of respiration: their function and normal tonus.
5. The causes of the excitability of the centre of respiration.
6. Apnœa.

THE DIAPHRAGM AND THE PHRENIC NERVES.

Genesis of Simple Respiration.

In the human subject respiration takes place, as is well known, by the diaphragm and by the muscles of the thorax, and these generally act at the same time. Still, under certain circumstances (as, for instance, during the sleep produced by chloral), according to Mosso, the want of agreement between the movements of respiration can become so great that abdomen and thorax may move in opposite directions for a considerable time. The same may be seen under certain circumstances in the rabbit. This we will discuss later. Also, the arrest of one or other system—thoracic or abdominal—of respiratory movements in the human subject, although generally accompanied by great disturbances of respiration, is not absolutely fatal, and can even be borne for a considerable time. We know that in the normal condition women breathe mainly by the thorax, men by the diaphragm; and Mosso has observed the interesting fact, that during sleep in both sexes respiration is carried on by the thorax, and that the diaphragm often moves so seldom that one might imagine this muscle had been paralysed. Charles Bell, Romberg, and others, have described the respiration of people who had sustained injuries to the vertebral column in the region of the lowest cervical and upper thoracic vertebræ, and in these people the intercostal and abdominal muscles of respiration were paralysed; only the diaphragm, serratus, sterno-cleido-mastoid, and trapezius performed still their functions. Inspiration, in these circumstances, consisted of short, quick breaths; expiration, and all actions connected therewith, were incomplete, being accomplished not by the contraction of muscles, but by the elasticity of the ribs and integuments of the abdomen, as well as by the pressure of the abdominal viscera against the under surface of the relaxed diaphragm. Similar symptoms are seen occasionally in spondylarthrocace of the vertebræ of the neck, when they last for a considerable time, and disappear with the cure of

the disease. Duchenne has observed in the living subject paralysis
of all the muscles of respiration, with the exception of the diaphragm,—
for instance, after a fall from a great height and injury of the
vertebral column at the level of the brachial plexus, and also by dis-
organization of the intercostal muscles caused by progressive atrophy.
In these cases, also, respiratory movement took place only in the
lower parts of the thorax and in the abdomen, however great were
the exertions made by the patient to distend his thorax so as to
inspire deeply. On the other hand, Duchenne has observed cases in
which complete paralysis and atrophy of the diaphragm existed to
such a degree that during inspiration epigastrium and hypochondrium
fell in, while the thorax expanded. Quite the opposite movement
occurred during expiration. "The patients," says Duchenne, "appeared
to aspirate their abdominal viscera by dilating their thorax during
inspiration." Such cases he observed in patients suffering from
hysteria, lead poisoning, and progressive muscle-atrophy, &c., and the
phenomena were observed sometimes for several years. In such cases
respiration takes place oftener than under normal circumstances,
appears to be only very slightly disturbed during rest and in sleep,
and is accomplished without the aid of any of the auxiliary muscles.
The inspirations take place mainly by the actions of the intercostal
muscles and the scaleni. But with increased exertion, extreme
dyspnœa sets in, and there is then great danger of choking. With
injury to the spinal cord above the origin of the phrenic nerves, death
generally takes place very quickly. Bell describes a person suffering
from such an accident, who lived for half an hour. The movements
of respiration were accomplished by means of the muscles of the neck
and shoulder only. With each inspiration the head was pulled
between the shoulder-blades. The diaphragm did not move. It
appears that long-continued contraction of the diaphragm is much
more dangerous than isolated paralysis of it. Duchenne mentions
in his famous book, *De l'électrisation localisée,* two such cases—the
one in consequence of rheumatism, the other during an attack of
tetanus. The second patient soon recovered after cessation of the
tetanus, while the first died in a short time from asphyxia.

Fick, supported by "geometrical considerations" and by personal
observations, regards the diaphragm, in the normal condition, as a

quite unimportant muscle of respiration in the human subject. Its innervation during respiration is, according to him, only to prevent it being sucked up. He considers the external intercostals as the most important muscles of inspiration. He also holds that normal expiration is active, and the abdominal muscles are therefore not necessary, but only the internal intercostals.

Further, in animals which generally breathe by thoracic movements, if both the phrenics be cut we find that more or less important disturbances follow. Panizza divided in dogs both the phrenic nerves and found that thereby respiration was quickened, the cavity of the chest widened to a greater extent and the muscles afterwards became more strongly developed, so that the cross section of the thoracic cavity increased. I can also confirm the statement that dogs survive the removal of both phrenics. Panizza found that after the operation in horses respiration was carried on with difficulty for several days and with somewhat powerful expiration. According to the same investigator, sheep die within two days after division of the phrenic nerves. The carcasses showed extensive venous congestion of the lower part of the abdomen, and there were numerous remains of food and gases in the stomach, contraction of the thorax (?), much blood and very little gas in the lungs.

In those animals which breathe mainly or exclusively by means of the diaphragm, the removal of the phrenics has much more important results. It is remarkable, however, that rabbits, although they breathe exclusively by means of the abdomen, under certain circumstances can live for months after division of the diaphragm nerves (the phrenics). Valentin observed this as early as the year 1848, and Budge confirmed the observation in 1855, while Arnold in 1836 recorded that respiration, after the phrenic operation, was only feeble and soon ceased altogether. In an account of a series of investigations undertaken, at a later date, by Budge in company with Eulenkamp, they described the extreme dyspnœa which appears in rabbits after the phrenics with all their roots had been cut, and they added that such animals died shortly after the operation. It appears that in consequence of this information some of the physiological text-books (see *Landois*) have taken up the erroneous notion that injury to both phrenics is fatal. Panizza reported (1865) that rabbits after division of both phrenic nerves died on the second day. In order to explain this contradiction, in the year 1879, I made, in company with Hugo Kronecker, a series of experiments on rabbits of different ages. In order to be sure that the phrenic is found and cut below the roots and the branches coming to it from the neck, the following is the mode of operation for the rabbit:—Make a median incision from the lower third of the neck to the upper end of the sternum, then pass along the mesial border of the sterno-mastoid, between this and the sterno-thyroid; here you will find the external jugular vein, and to the outer side of this, but more deeply placed, there is the cervical plexus from the 5th to the 7th cervical nerves. From the latter, going towards the middle, the phrenic nerve passes from without inwards over the scalenus muscle, accompanied by a small vein. Prepare the nerve from this spot a little further down, till immediately near the point where it passes beneath the subclavian vein and then into the cavity of the chest; at this place it may be easily and certainly divided, after it has received all its branches, which are a little inconstant in their origin.

When a phrenic is divided, the diaphragm on the corresponding side stops immediately, or the animal continues to breathe quietly, and there is no apparent distress of breathing. Immediately after division of *both* phrenics in older rabbits (above six months), although before the operation they had breathed quite quietly, there appears the most extreme dyspnœa. Respiration immediately becomes purely thoracic; with each inspiration the abdomen falls in, and it is pushed out with every expiration; whilst, on the contrary, with normal diaphragmatic respiration the abdominal integuments are pushed outwards with inspiration, and pulled in with expiration. All the respiratory muscles of the thorax, of the neck, and of the face take part; the nares become widely distended, even general convulsions of the body set in, and the symptoms are so threatening that the animal must be at once set free. When liberated, it lays itself flat on the abdomen, as if it sought a firm support for the contraction of its abdominal muscles. Gradually the animal quietens down a little, *it learns thoracic respiration*, and although the breathing continues to be strongly dyspnoic, with widely distended nostrils, the animal may live for months, its health being otherwise undisturbed.

(One of the animals which had been so operated on brought forth young two and a half months after division of the phrenic nerves.) This is only the case when older animals are operated on. Young rabbits, on the other hand, up to four months old, and even over that age, do not survive division of the phrenic nerves on both sides, but die, according to age, in a few minutes after the operation, showing strong convulsions and cyanosis, or in two or three days they die from asphyxia. When opened, the veins and right heart are found to be overfilled with blood, the lung is of a dark bluish-red hue, covered with numerous ecchymosed spots on the upper surface and on the inside.

It is now necessary to get more exact information regarding the changes in diaphragmatic respiration during and after division of the phrenics. In order to obtain a true picture of the changing form, height, and rapidity of the same, I used the graphic method by means of the Diaphragm-Double-Lever constructed by Kronecker and myself, and already mentioned in the report of the year 1879. This small and simple apparatus, on the same principle as that used by Rosenthal, but of much less complicated construction, gave in every respect such exact and satisfactory results that I used it in all my experiments for the registration of respiration in rabbits, and I can strongly recommend this method for the above purpose, especially as, with careful application, the animal may live without disturbance. The accompanying drawing shows the apparatus and the arrangement of the experiment. (Fig. 1.) The scoop-shaped end x of the small double lever Z can be quite easily, and without injury to parts, passed through a very small opening in the integument of the abdomen into the abdominal cavity. This can be best accomplished towards the right of the xiphoid process in the opening between this and the attachment of the last true rib, on the right side, to the sternum, so that when the operation is made, by pushing back the skin, no air can enter along with it. The scoop, by means of a very careful movement, is carried over the liver towards the diaphragm, and lies against the latter with its broad surface in opposition with the diaphragm, and the convex surface of the liver prevents it from springing off. On the upper arm of the lever is a movable knob b, on which there is a little hook. The latter connects a silk thread (a) with a long recording-lever, S. In order that the lever (S) should write on a smoked cylinder (K) which rotates on a horizontal axis with well-regulated friction, it moves in a Charnier joint, the axis of which is allowed to lean more or less forwards by means of a screw (*f*). Further, to bring the diaphragm-lever back to its position of rest after the contraction of the diaphragm has ceased,

Fig. 1.—Diaphragm-lever, Recording-lever, Chronograph, and Kymograph Cylinder arranged in the position for conducting the experiment. *Z*, Diaphragm-lever; *x*, scoop-shaped end of the same; *g*, Fork; *b*, movable knob of the same; *R*, Chronograph; *S*, Recording-lever; *f*, movable screw of the same; *K*, Kymograph cylinder; *r*, Wheel; *a a'*, Silk threads; *d d'*, Conducting wires to the Clock and Battery.

Fig. 1a.— Minute construction of the Recording-lever *S* with movable screw *f*.

on the other side, a silk thread (*a'*) is tied to the recording-lever, and this hangs over a vertically-placed wheel (*r*) by means of a small weight. The small fork *g*, which holds the rotating axis of the diaphragm-lever and which comes to lie close over the abdominal integuments, is screwed on to the cross-bar of a support, which stands near the rabbit-holder; this arrangement has proved to be more practicable than to attach the fork to the animal itself or to sew it on to the ensiform process, as slight displacements of the latter cannot be avoided. The length of the diaphragm-lever used by me was 125 mm; its axis lay 90 mm. from the apex. The lower (scoop-shaped) arm of lever is 35 mm. long and in rabbits reaches exactly to the middle of the diaphragm. The scoop at the end of the lever is 18 mm. long, at its widest part 6 mm. broad. When the apparatus is well

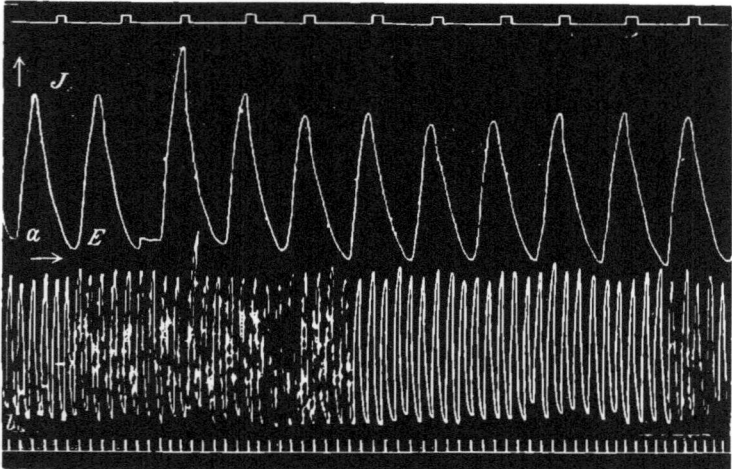

Fig. 2.—Normal diaphragm respirations of a rabbit. *a*, With quick movement of the drum; *b*, with slow movement of the drum. (The curves all run from left to right; the inspirations in diaphragm-respiration from below upwards.) *J*, Inspiration; *E*, Expiration.

arranged, then both the silk cords *a*, *a'*, at both sides of the recording-lever must form a horizontal line, while the small glass pen fixed to the end of the recording-lever registers on the horizontal drum of the Baltzar-Kymograph the curves of respiration. If during inspiration the diaphragm moves downwards, then the upper arm of the double lever will be moved backwards towards the thorax of the animal, and the registering lever follows in the same way. Accordingly, when the paper is taken off the cylinder it shows respiratory curves of the kind of which Fig. 2 is a facsimile. In this figure, the movement of inspiration is directed upwards, and the movement of expiration downwards. In order to judge of the method, I will insert here some of the curves:

Fig. 2 shows the diaphragm-respiration of a strong rabbit which has just been bound down and is tolerably restless, as follows: (*a*) with quick and (*b*) with slow movement of the drum.   Where there are no special remarks, the curves are to be read from left to right; when no

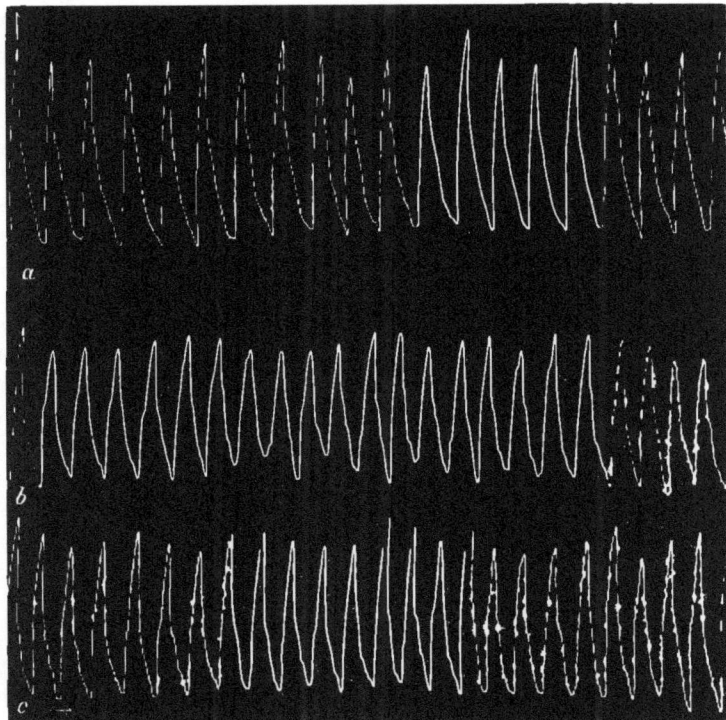

Fig. 3.—Diaphragm-respiration of a rabbit, in the trachea of which a Gad's T-canula is fastened. *a*, Nasal respiration; *b*, Tracheal respiration; *c*, when both ways are free.

"second" marks of time are given, the rapidity of the rotating cylinder-covering is the same as the rate of movement of the drum given in Fig. 2, *a*.

    In Fig. 3 the diaphragm-respiration of a rabbit is shown after tracheotomy and insertion of Gad's tracheal canula, viz.: (*a*) arranged so that animal breathed through the nose, the tracheal opening being closed; (b) arranged so that animal breathed through opening in neck, the nose being closed; (c) arranged so that both tracheal and nasal openings were free. From the last curves it will be seen, that with nasal arrangement of the canula the frequence of respiration is at its lowest; on the other hand the height of the individual respirations is at its greatest. The number of respirations in one minute was 72, the duration or length of a breath 0·81″. The height of a respiratory curve varied from 2·7 to 3·1 cm. When both ways stood open, then respiration was quickest: 90 in one minute; length of one breath 0·75″, and height of a single respiration 2·1 to 2·8. Through the tracheal opening of the canula, the number of respirations was 85 in one minute, length of a breath 0·75″, height of a single respiration 1·6 to 2·5,—on the whole very similar to the observations made when both ways stood open. This shows that the rabbit had its necessity for respiration completely satisfied through the opening of the trachea, an important fact in connection with later experiments. After the phrenic nerve had been prepared on both sides, laid on threads, and the animal had been quiet for some time, with nasal arrangement of the canula, the number of respirations was 54 in one minute, length of a breath 1·06″, and height of a single respiration 4·4 to 5·1 cm., showing an irritable condition of the phrenics. With both ways open, 61 per minute, length of one breath 0·75 to 0·8″, and height 3·8 to 4·8 cm. With tracheal breathing, 60 in one minute, breath 0·81″, and height 4·0 to 5 cm. The respiration thus became much slower and deeper compared with the foregoing.

If a phrenic nerve be now cut, in the way already mentioned, the diaphragm stands still on the corresponding side; on the other side, it breathes away quite undisturbed, scarcely with greater frequence and without the thorax taking part. On the other hand, after division of the second phrenic, the picture changes as if by magic (Fig. 4). The diaphragm remains for a few seconds under oscillating movements in a completely relaxed condition; then strong general movements of the muscles appear, by which the animal shakes very much, whereon the diaphragm passively follows the movements of the thorax which are now beginning, and it is therefore drawn into the thorax by inspiration, and pushed downwards by expiration. The upper arm of the diaphragm-lever accordingly is moved forwards by inspiration (towards the abdomen of the animal), backwards by expiration, so that now the registering point notes the inspiration

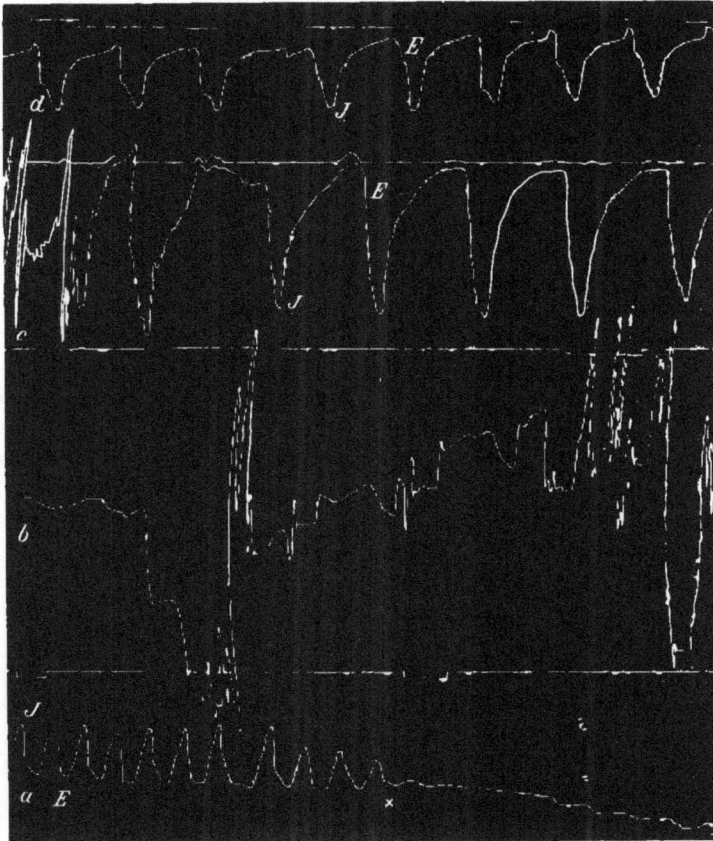

Fig. 4.—(To be read from below upwards.) Respiration of a rabbit before, during, and after division of both Phrenics. *a*, Normal respiration; at X, division of both Phrenics; *b*, irregular respiration after division of the Phrenics; *c* and *d*, thoracic respiration; at *c* strongly dyspnoic; at *d* a little quieter (at *c* and *d* the thoracic inspirations run from above downwards, because the diaphragm is drawn into the thorax by the inspirations). *J*, Inspiration; *E*, Expiration.

stroke in the opposite direction. In Fig. 4, *c* and *d*, therefore, the direction of inspiration is from above downwards, and of expiration from below upwards. At the same time respiration becomes much slower, often by a half and more, while the duration of a single breath very often is doubled. In this way, respirations sank in number, for instance from 85 to 42 in one minute; and at the same time the length of a breath rose from 0·75 to 1·5″; in another case, the number of respirations diminished from 56 to 20, while the duration of a respiration rose from 1·1″ to 3″; in a third case, before division of phrenics, the rate of respiration was 62 in a minute; after the operation, 42 in one minute; and the length of a breath rose from 1″ to 1·8″. In Fig. 4 we have a typical picture of respiration movements after division of the phrenic nerves on both sides; the four curves, one over the other, form a regular series; we recognize on curve *a* the stoppage of the diaphragm; on curve *b* the irregular movements of the muscles of the abdomen and thorax, and on *c* and *d* the pure thoracic breathing.

The phrenics, however (as Budge and later Panizza found), contained sensory fibres, so when they were cut the animals gave signs of pain. The great restlessness and movements of the whole body seen after division are probably mostly caused by this, so that often one is not successful in obtaining tracings of the first respirations after division or ligature of the phrenics. In order to avoid this restlessness at the beginning, it is recommended to break the nerve connection by means of sudden cooling after Gad's method. For this purpose, I have

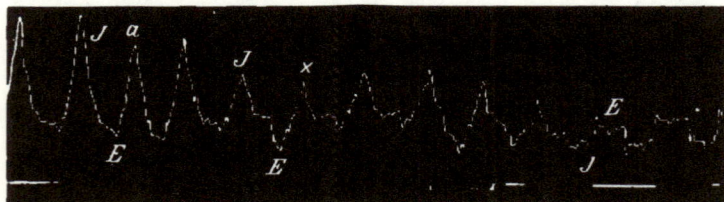

Fig. 5.—Cooling of the Phrenics according to Gad. At *a* beginning of cooling; at *x* transition of the diaphragm respiration into thoracic respiration. *J*, Inspiration; *E*, expiration.

laid two small silver tubes under the phrenics, and I have allowed a freezing solution giving a temperature of −5° C. to flow through them. The sensory fibres are paralysed so quickly by this that the animal remains quiet and one can now register exactly the movements from the beginning of the suspension of the action of the motor fibres of the phrenic nerves thus caused by cooling. The effect on the diaphragm is the same as after division, with this difference, that the stoppage of the diaphragm does not take place so suddenly, but by contractions becoming gradually smaller. Fig. 5 shows the effect of cooling the phrenic nerves, in which one sees the gradual transi-

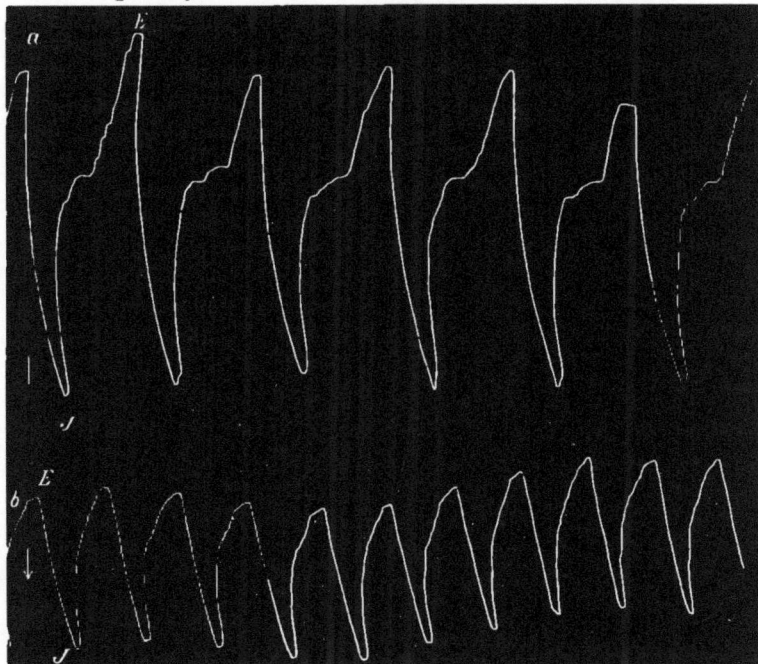

Fig. 6. · Thoracic respiration of a rabbit several days after division of Phrenics on both sides. (a) Respiration through the nose; (b) respiration through the tracheal canula.

tion from diaphragmatic to thoracic respiration. We have observed, that when in rabbits thoracic respiration appears after division of the phrenic nerves (which under normal condition is latent), respiration becomes much slower, deeper, and at the same time dyspnoic. Thoracic respiration is therefore hardly sufficient to replace abdominal respiration. This dyspnoic thoracic respiration may be noticed for weeks after the operation; the replacement will be somewhat more complete when a tracheal fistula has been made, which diminishes the opposition to the inspiration stream. Fig. 6 shows the thoracic respiration of a rabbit in which the phrenic nerves had been divided several days before the tracing was taken; at a the animal breathes through the nose; at b through the tracheal canula.

But in young rabbits the thorax is very soft, and can be pressed in, and its muscles seem inadequate for the great work which is necessary to sufficiently increase the thoracic capacity, the latter having been diminished by the diaphragm being relaxed and sucked up. Probably herein lies the reason why young animals die so quickly after division of the phrenics.

In order to show that distress of breathing is brought on by want of air, we only require to measure the volume of air which rabbits inspire in a given time before and after removal of the influence of the phrenics. For this purpose, I allowed the animals to inspire the air from a spirometer, and to expire in the free atmosphere. This was done after the method of Rosenthal, who did the same thing, but for other purposes. By means of a simple electric arrangement, the quantity of air taken in was noted on the kymograph-drum under the respiratory curves. The arrangement is as follows (Fig. 7):—The stem of a glass T-tube (t) is attached to the air-cock (K) of a Hutchinson's spirometer (S) by means of an india-rubber tube. In this india-rubber tube is fastened a Voit's (pear-shaped) valve (v), which is shut by water. This is brought into the position illustrated in Fig. 7 (by v). The water surface is so placed that the air on its way from the spirometer to the animal finds very little resistance. Air passing in the opposite direction (expiration) stops the valve, by lifting the column of water. One end of the horizontal limb of the T-tube is in air-tight connection with the tracheal opening of Gad's canula, while at the other end there is an expiration valve (e). This valve consists of a piece of moist gold-beater's skin, which lies lightly on the somewhat puffed-out edge of the short expiration tube. With the inspiration of the animal, the atmospheric pressure pushes this membrane into the glass tube and prevents the entrance of air, so that the animal can obtain its air of inspiration only through the inspiration valve from the spirometer. Through the expiration pressure, on the other hand, the gold-beater's skin is lifted, and allows the air to escape, while the inspiration tube is closed by means of the water-valve. In order to

register electro-magnetically the volume of air taken from the spirometer, the arrangement is made that the pointer (Z) attached to the
sinking air-cylinder (C) always causes a short closure of an electric

Fig. 7.—Spirometer respiration of a rabbit. S, Spirometer; C, Air-cylinder; K, Air-cock of the same;
q q', India-rubber tubes filled with mercury; h h', Cocks for filling the same; Z, Brass pointer of the air-
cylinder; v, Pettenkofer's Inspiration Valve; e, Expiration Valve (gold-beater's skin); t, Glass T-canula;
g, Gad's Canula; a a' a", India-rubber tubes; z, Diaphragm lever.

current when 200 ccm. of air have passed out of the spirometer. This is accomplished by means of the following simple arrangement; The hard vulcanite scale of the spirometer has brass plates fixed at regular distances, and the space between any two of these plates represents 200 ccm. of air. These brass plates are punched through the vulcanite, and on the opposite side of the scale they are attached to one another by means of a wire. The metal pointer of the air-cylinder, which, by the sinking of the latter, glides along the hard vulcanite scale, is connected with one pole of a Daniell's element; while the connection of the other pole, through an electro-magnetic marker, runs to the spirometer scale. Each time that the pointer with the sinking cylinder touches a brass plate the circuit is closed, and the writing magnetic lever notes that 200 ccm. have been taken away. In the circuit of the spirometer scale and of the writing magnetic lever, a clock, registering seconds, is placed; and in this way we obtain, in a line above the respiration curves, the data for the calculation of the time during which 200 ccm. of air are breathed by the rabbit.

In order to prevent the resistance increasing, by the sinking of the air-cylinder into the water-cylinder (which the animal has to over-come in order to obtain air), firstly, the air-cylinder is made heavy with weights, so that a minimum overpressure is sufficient to cause the cylinder to sink; secondly, an arrangement is made, analogous to one used by Mosso for his first plethysmographic investigations, so that the weight on the cylinder increases as it sinks. This is necessary, because the weight of the floating cylinder on the air becomes smaller the deeper it sinks into the water. The way in which the loading is gradually increased is effected thus: The cords which, conducted over wheels, support the cylinder are covered with india-rubber tubes filled with mercury. It is quite easy to give the mercury column the necessary diameter, so that their increas-ing length, whilst the cylinder sinks, increases the weight of the floating cylinder as much as the increased water-level diminishes it. This arrangement had the desired effect, as I shall immediately prove by means of figures; so that, as the animal was allowed, by means of Gad's canula, to take the air either from the free atmos-phere through the nose, or through the tracheal opening out of the spirometer, the animal preferred the easiest way, and took more air from the spirometer than from the atmosphere. In order to attain accuracy, I took care that the experiment should be discontinued when the air-cylinder was half emptied. On reaching this point, I began the experiment again with the cylinder quite full.

In this way, I completed a whole series of experiments, the results of which are given clearly in the following tables. The first two tables are only examples to show that respiration from the spirometer does not cause any special difficulties. Table I. is obtained from the same rabbit whose respiratory curves are registered on Fig. 3.

TABLE I.

	RESPIRATION WITH GAD'S CANULE FROM THE FREE AIR.			RESPIRATION WITH GAD'S CANULE FROM SPIROMETER.	
	Nasal Arrangement of the Canule.	Tracheal Arrangement of the Canule.	Both ways free.	From Spirometer alone.	From Spirometer and Free Air.
Number of respirations in one minute	72	85	90	76	90
Length of a single breath	0·81″	0·75″	0·75″	0·75 – 0·81″	0·68″
Height of a single respiratory movement	2·5 – 3·5ᶜᵐ	1·7 – 2·5ᶜᵐ	1·8 – 2·7ᶜᵐ	3·1 – 4·1ᶜᵐ	3·1 – 4·1ᶜᵐ
Quantity of air in one minute	—	—	—	906ᶜᶜᵐ	766ᶜᵐ
Quantity of air with each respiration	—	—	—	12ᶜᵐ	8·5ᶜᵐ

Taken from Spirometer.

TABLE II.

	RESPIRATION WITH GAD'S CANULE FROM THE FREE AIR.			RESPIRATION WITH GAD'S CANULE FROM SPIROMETER.	
	Nasal Arrangement of the Canule.	Tracheal Arrangement of the Canule.	Both ways free.	From Spirometer alone.	From Spirometer and Free Air.
Number of respirations in one minute	51	56	48	52	52
Length of a single breath	1·1″	1·1″	1·1″	1″	1·1″
Height of a single respiratory movement	3·2 – 4·0ᶜᵐ	2·1 – 2·7ᶜᵐ	3·0 – 3·4ᶜᵐ	2·3ᶜᵐ	2·8 – 3·1ᶜᵐ
Quantity of air in one minute	—	—	—	614ᶜᵐ	325ᶜᵐ
Quantity of air with each respiration	—	—	—	11·8ᶜᵐ	6·25ᶜᵐ

Taken from Spirometer.

Table III. shows how much air rabbits at different ages inspire in a given time before and after division of the phrenics. One sees plainly, that while old rabbits, after removal of the influence of the phrenics, can still reach from $\frac{1}{2}$ to $\frac{8}{9}$ of the normal quantity of air used, in young animals the proportion stands so that they only respire from $\frac{1}{4}$ to $\frac{1}{6}$ in one minute; and thus they satisfy their need of air so unsatisfactorily that they soon die.

TABLE III.

Cases.	SPIROMETER RESPIRATION BEFORE DIVISION OF PHRENICS.			SPIROMETER RESPIRATION AFTER DIVISION OF PHRENICS.					
	Number of Respirations in 1 Min.	Quantity of Air in 1 Min.	Quantity of Air with each Respiration.	Number of Respirations in 1 Minute.		Quantity of Air in 1 Min.		Quantity of Air with each Respiration.	
				Immediately.	Later.	Immediately.	Later.	Immediately.	Later.
		ccm.	ccm.			ccm.	ccm.	ccm.	ccm.
Older Rabbit	76	906	12·0	42	60	220	425	5·2	7·0
Old Rabbit	60	546	9·1	41	51	360	429	8·7	8·2
Old Rabbit	57	747	13·1	47	51	370	459	7·9	9·0
Very Old Rabbit	79	966	12·2	67	72	728	847	10·9	11·8
Old Rabbit	49	533	10·9	45	—	344	—	7·7	—
Young Rabbit	56	614	11·8	20	23	150ccm		6·5ccm	
Very Young Rabbit	72	854	11·8	30	—	130ccm		4·4ccm	

If, instead of dividing the phrenics, one manages to stop their action by means of cold, then, if the degree of cold has not exceeded $-2°$ C., by heating the phrenics the nerves soon regain their power of action (this is accomplished by passing water at $37°$ C. through the tubes); and in this way one can work out the effect of removing the influence of the phrenics several times in succession. At $-5°$ C. the nerves are killed for ever. Table IV. gives an example of an older rabbit, in which repeated cooling and heating of the phrenics had taken place.

Following immediately on repeated cooling and heating, division of the phrenics does not alter in any way the results.

From the experiments regarding division of phrenics, as already mentioned, we learn that the length of a respiratory phase diminishes when the abdominal type of breathing merges into the thoracic. Thus, while before division of the phrenics the length of a phase varied between 0·75″ and 1·5″, it increased after division to two, and at times to three, seconds.

TABLE IV.—OLDER RABBIT.

SPIROMETER RESPIRATION AFTER HEATING OF PHRENICS.				SPIROMETER RESPIRATION AFTER COOLING OF PHRENICS.			
State.	No. of Respirations in one min.	Quantity of air in one minute.	Quantity of air with each Respiration.	State.	No. of Respirations in one min.	Quantity of air in one minute.	Quantity of air with each Respiration.
Before Cooling }	49	ccm. 533·0	ccm. 10·9	After Cooling }	45	ccm. 344·8	ccm. 7·7
After Warming }	49	590·4	12·2	After Cooling }	43	362·0	8·4
After Warming }	49	571·8	12·4	After Cooling }	42	358·0	8·5

Nature of a diaphragmatic contraction.—With reference to the doubt, still often expressed, regarding the discontinuity of voluntary tetanus and tetanus caused by reflex action, it seemed proper to decide whether the simple respiratory movement of the diaphragm was to be regarded as a quivering or as a short tetanus. For this reason, the synthesis and analysis of contraction were investigated.

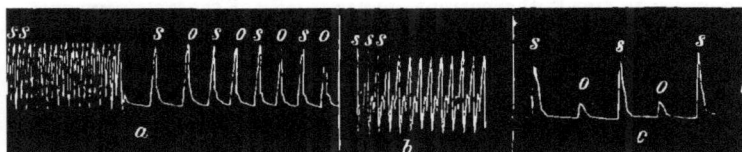

Fig. 8.—Single quiverings of the diaphragm. *S*, Closing; *O*, Opening induction shock; *a*, maximal; *b* and *c*, under maximal strength.

For the first purpose, both phrenics were stimulated by single induction shocks. It was shown that the duration of the quiver did not alter to any appreciable extent, neither by closing nor opening induction shocks, nor by quick moving in and out of the secondary coil of the induction machine, nor by insertion or removal of resistance in the circuit of a galvanic current. The time amounted in the

fresh experiment to from 0·125″ to 0·3″; and in the case where the
nerve had been already used, to 0·5″, also from $\frac{1}{2}$ to $\frac{1}{3}$ of a normal
respiration. Fig. 8 shows such single quivers of the diaphragm.
The period of latent stimulation in the unexhausted muscle lasted
0·035″. Fig. 9 shows the curve of a simple quiver of the diaphragm

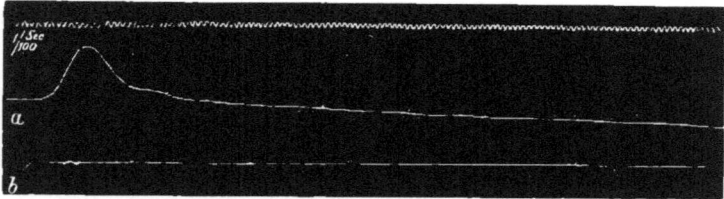

Fig. 9.—Simple contraction or twitch of the diaphragm. *a*, Curve of the diaphragm contraction;
b, Shock marking line, showing moment of stimulation at ascent of curve a little to right of *b*.

at the quickest rate of movement of the drum. When, during
this single stimulation, the thoracic respiration continues, then the
quiver curves appear as pointed teeth on the complete respiratory
curve. Tetanizing stimulation of moderate intensity overcame natural
respiration. We now attempted through rhythmic tetanization of the
peripheral ends of both phrenics to cause a mode of respiration analogous to the natural one, so that we might find out the frequency of
stimulations necessary to produce respiratory curves like the normal.
In order to break the current intermittently, Kronecker's magneto-
electric-vibration apparatus, with wash-contact, after removal of the
Wagner-hammer, was connected with Du Bois-Reymond's sledge
inductorium (the common form of induction coil used by physiologists).
When we now stimulated the peripheral ends of the phrenics at inter-
vals of from $\frac{1}{10}$ to $\frac{1}{15}$″, the diaphragm lever registered teeth quite
distinctly on the inspiration summit of the respiratory phase. Only
when about twenty stimulations in the second were sent through the
phrenics did the teeth become invisible (Fig. 10), and we obtained
respiratory curves quite similar to the normal ones when we stimu-
lated rhythmically every second for a period of $\frac{1}{6}$ to $\frac{1}{3}$ of a second
each time. For the latter purpose, a Baltzar stimulating clock was so
arranged by a double interrupter being interpolated in the circuit of
the secondary spiral of an induction apparatus, that in every second

for a period of $\frac{1}{4}''$ or $\frac{1}{3}''$ the contact was closed, and for the rest of the second was opened.

If we now divided the *medulla oblongata* below the centre of respiration so that each individual respiration was arrested, by direct

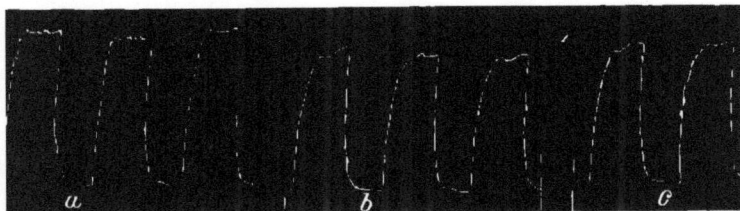

Fig. 10 —Artificial electric diaphragm-respiration by means of rhythmic stimulation of the peripheral phrenics with intermittent shocks, after separation of the medulla below the respiratory centre. *a*, Stimulation with $\frac{1}{15}''$ interval; *b*, $\frac{1}{15}''$ interval; *c*, $\frac{1}{5}''$ interval. Distance between coils of induction machine 20 mm.; 1 Daniell.

electric stimulation of the phrenics, an artificial respiration could be induced, which saved the animal from asphyxia for any wished time. I have kept it up for half an hour.

This raised a fundamental question regarding the transmission of a stimulus from the central nervous system: How many and what kind of stimuli are necessary to cause the respiratory centre to discharge for *one* respiration? This question we will come back to later on.

II.

SPINAL CORD TRACTS OF RESPIRATION. SPINAL RESPIRATORY CENTRES.

It is scarcely thirty years since we began to get some exact ideas regarding the microscopical structure and physiological functions of the spinal cord. Still, in the year 1842, Budge, backed by experiments, thought he was right in coming to the conclusion that the phrenic nerves pass up through the spinal cord "and end in the *medulla oblongata*, next to the other nerves which innervate the parts of the body moved involuntarily," and that they could be directly stimulated at that spot. He had also observed on stimulation of the *corpora striata*, of the *thalami optici* or of the *corpora quadrigemina*, with a needle which had been dipped in an acid,—immediately after the death of the animal, when all voluntary diaphragm movements had disappeared,—that after a few seconds strong contractions of the diaphragm and lower intercostal muscles took place. When he repeated these experiments on living rabbits and cats, he came to the conclusion that respiration was quickened, and that in the rabbit it rose from 64 to 89 in the minute. Higher parts of the brain did not show any such relation to respiratory movements. Long before Budge, Haller, and later Flourens (1824), attempted a direct irritation of the *medulla cervicalis*, divided above the origin of the phrenic nerves, but they only obtained slight respiratory movements. Budge, in the year 1855, renewed his experiments, at a time when it was already known that the motor nerve-roots leave the spinal cord at the same level as that at which they spring from the anterior cornua of the gray matter; but he did not now succeed, by means of electric stimulation of the *medulla oblongata* after the death of the animal, in obtaining any effect on respiration movements. He was obliged to carry the electrodes to the level where the roots of the third and fourth cervical nerves emerge from the spinal cord "in order to obtain a movement of the diaphragm for a given time, similar to the normal movements of respiration." Budge now concluded that the phrenic nerves do not ascend to the *medulla oblongata*, but that stimulating fibres from the latter descend to the *medulla spinalis*, and that these fibres could not be stimulated by galvanism. ("*Nec irritationis galvanicæ capaces esse.*") Since then, direct stimulations of the *medulla spinalis* have often been tried.

It is now generally acknowledged that by stimulation of the sensory tracts of the spinal cord, motor effects may be obtained through reflex action. Still, opinions regarding the direct excitability of the so-called motor spinal cord tract are much divided. While Fick and Engelcken, later Fick alone, Birge, then Luchsinger, and especially Biedermann, and in more recent times Mendelssohn, by their experiments, were led to believe in the direct stimulation of the motor spinal cord tract; to-day Schiff still agrees completely with the conclusions arrived at by Van Deen, that the motor portions of the spinal cord in mammals are not excitable by direct stimulation, but are "kinosodic," that is, where movements appear, the same are occasioned by stimulations exciting reflex actions. Whatever may be the interpretation of Biedermann's experiments, their accuracy completely confirms Schiff's view. For myself, the important question now arose for investigation, whether the *medulla spinalis* receives stimulations, which are followed by movements of the diaphragm, after the *medulla oblongata* has been cut off from the spinal cord, and if it does, of what kind they may be?

Biedermann's research was made on the lower extremities of frogs, with the following remarkable results. The first result of direct stimulation of the front and partly lateral fibres of the spinal cord consists in a greater or less tetanic movement of all the muscles, which sometimes passes into complete tetanus when the coils of the inductorium are brought close to each other. Often co-ordinate movements take place. Similar results can be obtained from a mechanical stimulus as from an electrical stimulus (cutting through slowly, tying the spinal cord gradually tighter). Narcosis, loss of blood, asphyxia, as well as the state of shock or collapse into which the spinal cord of some of the higher vertebrate animals falls for a considerable time after severe injuries, (and in which, according to experience, reflex functions are apparently extremely weakened), also influence the result of direct stimulation. Single induction shocks have a stimulating effect only when they are very intense, whilst opening and closing shocks in quick succession, or the rapid opening and shutting of a battery with elements in series, have an effect if of comparatively weak intensity. This shows, says Biedermann, "that the ganglionic elements of the gray substance of the central nervous system possess the peculiarity in a high degree of summing up or fusing of ineffective stimulations. It may, however, be left undecided whether these stimulations cause a continued but latent state of excitation in the ganglionic elements or only leave a condition of higher excitability in them." The stimulating action increases gradually in power by means of rhythmic stimulations of a moment's duration (induction shocks) separated by comparatively short pauses. This occurs up to a certain point, and under similar circumstances its ineffective stimulations may gradually become effective. Further, it appears, according to Biedermann, that after the effect of a stimulation has passed away, the same currents (abterminal opening), which were absolutely ineffective before, now cause strong contractions, and that this influence only disappears gradually within a period of several seconds (*The Formed Path of Exner*), &c. We therefore learn from Biedermann's experiments: 1, That the first result of electric or mechanical stimulation of the anterior fibres of the spinal cord is a tetanic agitation of all the muscles, and this sometimes merges into true tetanus; 2, that thereby co-ordinate movements often appear; 3, that single induction shocks have a stimulating effect only when they are of great intensity, but that the stimulation increases in power when the shocks follow one another at not too long intervals, so that even much weaker shocks thus become effective; and 4, that rhythmic electric stimulation works in proportion to the aggregation or summation of the stimulations.

These results stand in full concord with the general laws which Stirling drew from his experiments on the reflex function of the spinal cord, and which partly read as follows:—"Comparatively strong single electric shocks of the skin are summed up even if they follow at an interval of two seconds. Shocks which quickly follow one another always induce a reflex action. With fatigue, the period of latent stimulation increases; still, ineffective shocks are often followed by an increase of excitability. The summation effect may continue after a weak liberation and cause stronger liberations. When the shocks are not very frequent, the time of summation can be greatly varied by altering their strength and frequency. Frequency and intensity can in this way replace one another, so that strong shocks coming at long intervals cause the same time of reflex action as weaker shocks coming at shorter intervals. Single induction shocks must be very strong in order to cause reflex movements in legs whose skin is irritated. The duration of the period of latent stimulation increases very quickly with fatigue, and soon the limb does not give any reaction with individual shocks."

Knowing that the results of Stirling and Biedermann agreed, I did not think it necessary to undertake an isolated stimulation of the anterior or posterior spinal cord tracts, but contented myself with stimulating the spinal cord as a whole, in order to study the result of this electric stimulation on the diaphragm. I have, in rabbits, and in exceptional cases in cats, transversely divided the *medulla oblongata* close below the centre of respiration in the fourth ventricle or a little deeper, close below the point of the *calamus scriptorius*, so that the animal could no longer breathe. Artificial respiration was at once caused by means of Kronecker's respiration apparatus, and in a manner already described by me in an earlier work, with this modification, that before the air reached the animal it was warmed. Thus the animal was protected from too rapid cooling, and could be maintained for a longer time under normal conditions than when it breathed cold air. Artificial respiration was practised till the stimulation of the spinal cord began; sometimes also it was continued during the time the stimulating process went on. The electrodes were two fine pearl needles, which were passed through a cork, varnished to the points. These were sunk into the substance of the cervical cord, immediately below the surface of the incision. When I stimulated the spinal cord with single opening or closing induction shocks, and used weak and moderate currents, I saw no effect on the diaphragm. It was only when the secondary coil was pushed almost completely over the primary that these appeared, accompanied by general restlessness of the extremities. Each opening and closing shock then caused a twitch or quiver of the diaphragm of known form, never a true respiratory movement of the diaphragm. The opening contraction was generally higher than the closing contraction. There was no doubt that these results were caused by the current stimulating the roots of the phrenic nerves. In one case, in which the *medulla oblongata* had been cut close below the respiratory centre, so that even after long artificial respiration no spontaneous respiratory movements appeared, strong induction shocks were able, even when they followed at long intervals of four

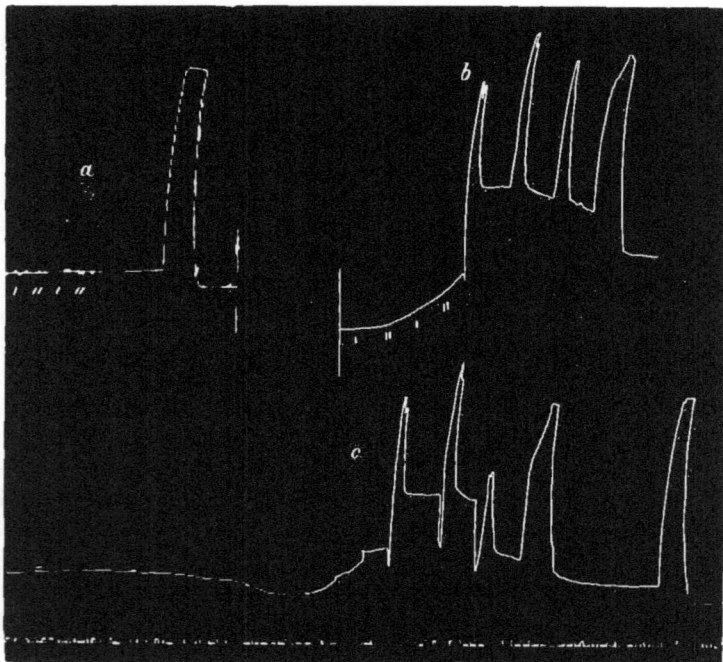

Fig. 11.—Respiratory spasms after stimulation of the spinal cord with strong induction shocks at intervals of 3 to 5 seconds in decapitated animals (below the *medulla oblongata*). *a*, Respiratory spasm of 4 seconds duration after 2 closing (') and 2 opening induction shocks ('') at intervals of 3 seconds, coils 80 mm. separate, 2 Daniells. *b*, Group of short respiratory spasms after 4 single shocks at intervals of 4 seconds, coils 80 mm. separate, 2 Daniells. *c*, Spasm group with resulting single spasm after 7 single shocks, at intervals of 4 to 5 seconds, coils 7 mm. apart, 2 Daniells.

seconds, to induce respiratory spasms by summed-up stimulations. Then very peculiar phenomena appeared, which give us a glimpse of the mechanism required for the accomplishment of respiration. This is illustrated by Fig. 11. We see (1), at *a*, a respiratory spasm having a duration of four seconds, which appeared after four opening and closing induction shocks, (at intervals of about three seconds,) had

Fig. 12. — Diaphragm spasm caused by direct stimulation of the spinal cord with intermittent currents (1500 Unit. 2 D. $\frac{1}{10}$ interr.). The curve runs from right to left. (Rabbit.)

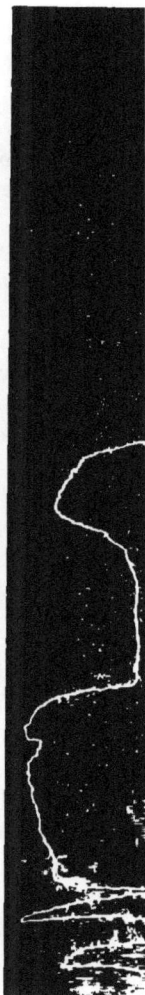

reached the *medulla cerviculis*, and eleven seconds had passed since the last shock; (2), at *b* a group of short respiratory spasms after four single shocks at an interval of about four seconds; (3), at *c* a spasm group followed by a single spasm after seven ineffective shocks at the long interval of from four to five seconds. At the same time, the curves show how the diaphragm gradually descended during the stimulations before reaching the level of tetanus, which represents the respiratory spasms. On the other hand, if the spinal cord was stimulated by intermittent shocks of $\frac{1}{12}''$ to $\frac{1}{20}''$ interval, especially at the latter rate (which in my experiments on the phrenic nerve I found to be the most favourable), and indeed either in pauses of 0·5 to 1 second, or for a longer time without pause; then we saw (after the stimulation had acted for some time, for instance, after fifteen seconds with moderate intensity of the current (500–700 Unit.)), along with clonic contractions, especially of the muscles of the body, first an inspiration spasm of the thorax appearing, and then, later, tetanus of the diaphragm. Then there generally appeared a smaller and shorter diaphragm spasm. The diaphragm rose quite gradually, sometimes to its maximum point of contraction, and then went back quite slowly to its original position of rest (Fig. 12). We not unfrequently saw—but only after a thoracic inspiration spasm had occurred—a row of diaphragm contractions follow one another and gradually increase in size; the next contraction setting in before the preceding one had disappeared, so that the whole proceeded like an avalanche. The duration of a respiratory period, as well as its height, varied greatly according to the duration and intensity of the current and also according to the excitability of the spinal cord. I have seen such from 34 seconds and 4·3 cm. in height to 8 seconds and 0·6 cm. in height. In Fig. 13 we see a curve of several contractions, increasing in size one after the other. The four periods in this case amounted to—

1. Seven seconds by 0·3 cm. height.
2. Nine   „   „ 1·1 cm.   „
3. Thirteen „   „ 2·7 cm.   „
4. Eleven   „   „ 3   cm.   „

The stimulation during the fourth period had to be stopped before the spasms ceased, as general spasms of the muscles appeared. These assumed a serious character, and demanded immediate recourse to artificial respiration. It is plain that such movements of the trunk and of the diaphragm have nothing in common with the normal respiratory movements of the rabbit. They are really respiratory spasms. The shortest spasm of this kind caused by direct stimulation of the *medulla spinalis* is still eight times longer than a normal respiratory movement of the rabbit, if we take the number of the respirations of a rabbit, which has been bound down for some time, at 60 per minute. Besides, the general movements of the muscles of the body which accompanied each breath pointed to the fact that they were of the nature of respiratory spasms. Here, of course, we have not to do with currents passing in the phrenic nerve. If we now move the electrodes in the direction of the spot where the nerves of the diaphragm originate, and if we stimulate there the *medulla spinalis* with intermittent shocks of $\frac{1}{30}$ second interval, then we can, according to the duration of the stimulation, make the movements of respiration long or short according as we wish, just as was done by direct stimulation of the peripheral phrenics. We have, therefore, in the spasms of the diaphragm to do with a summation mechanism, caused by rhythmic shocks, which, according to all researches down to the present time, is characteristic of the mode of action of the *medulla spinalis*, whether it be stimulated directly or by reflex action. At the same time, it was very instructive to notice that in rabbits (which in the normal condition only breathe by means of the diaphragm), when the spinal cord was directly stimulated, the respiratory muscles of the thorax, nevertheless, contracted first. This fact can only be explained by supposing that the roots for the respiratory nerves of the muscles of the thorax, lying nearer the electrodes, are first stimulated by the currents, and then the roots of the phrenics. In vertebrates from which the brain has been removed, stimulations coming from the skin, caused by pinching the paws, the tail, &c., only seldom, and under favourable conditions, liberate respiratory movements by reflex action.

Therefore, on the whole, my results regarding the diaphragm of the rabbit, by direct stimulation of the *medulla spinalis*, do not contradict the results which Biedermann, by direct stimulation, and Stirling by reflex stimulation of the spinal cord, saw on the muscles of the hinder extremity of the frog. I have seen no effects which indicated the existence of a respiratory centre in the spinal cord; on the contrary, everything seemed to point to the fact that we had to do with ordinary conducting tracts. Of course, and this is an important point, the stimulations of the conducting tracts of the spinal cord are not propagated as along motor nerves, *but as if they passed along sensory nerves*, which end in the masses of gray matter forming the centres for the phrenics. This is the reason why single shocks (which, in motor nerves, cause a contraction of the muscle), as a general rule, give no action; whilst intermittent shocks are summed up, and then the liberation of movement follows.

How do these results compare with the observations which led Langendorff to say that the true centre of respiration lies in the spinal cord, while the respiratory centre in the *medulla oblongata* is only a regulating centre, or an accessory or helping centre? Brown-Séquard, as early as 1855, made the remark that excision of the *nœud vital* in the *medulla oblongata* was not necessarily fatal. Then Rokitansky and V. Schroff, in 1874, in rabbits in which they had divided the spinal cord in the neck, and which they had poisoned with strychnine, saw respiratory movements during the strychnine spasms. These latter observers concluded that in the *medulla spinalis* there exists a centre for rhythmic movements, which in life is active only while in functional connection with the brain, but by means of strychnine works for a short time after removal of this connection. Finally, Langendorff drew special attention to the respiration of newly-born animals, "in which the *medulla oblongata* had been cut immediately (or a few millimeters) below the point of the *calamus scriptorius*," and which after that breathed automatically for a considerable time "like normal animals." He further observed that this was especially the case after a weak dose of strychnine had been injected. In such animals, during the respiratory pauses of artificial respiration, by gentle stroking and blowing on the skin, nipping the paws and tail, or by electric stimulation of the sciatic nerve, Langendorff found he could liberate strong respiratory movements; and, in fact, with one shock sometimes a whole series of them. He often obtained no reflex action, the fatigue of the centre, which had been stimulated so often in rapid succession, being very manifest. Then he was successful, by means of small doses of strychnine, in making the reflex actions immediately visible. He could also make the animals (having only their spinal cords) apnœic, and then all the reflex actions exciting respiration disappeared. The spinal centres of respiration, according to Langendorff, can be stimulated not only by reflexes, but they also work automatically—that is, they are not only reflex but also automatic. In newly-born animals he saw, during artificial respiration, or also during the time the latter was suspended, whole series of slow but quite regular respirations, such as appear after division of the vagus. These respirations soon became very weak, but they could be kept up by means of strychnine. Langendorff thinks he has convinced himself that these movements are really respirations and not muscle spasms; and also, that they are not the result of stimulations in consequence of the division of the *medulla oblongata*. The apparent or actual suspension of respiration after division of the *medulla oblongata* is to be regarded, according to Langendorff, as an effect of inhibition, and not, as thought by Goltz, Brown-Séquard, and others, a mere loss of function.

My own experiments on rabbits and cats, to which strychnine had been administered, have given me certain results when the animals were young, and under exceptional conditions even on older animals not poisoned, when they had been made so cold blooded that the heart, removed from the body, continued to beat for some time like the heart of a frog. Thus after *sectio bulbi*, spontaneous and reflex movements of the diaphragm may occur; but the tracings give evidence of what can already be learned by the eye: that in these cases we have to do, not with normal respiratory movements, but with *spasms* of the respiratory muscles. After division of the *medulla oblongata* the cold-blooded rabbits showed reflexes only for a very short time. One succeeded by pinching the skin or the tail in liberating a long diaphragm spasm, but during this spasm the entire hinder portion of the animal raised itself, if left free. This experiment could be twice or thrice repeated on the same animal with success. At the same time I observed in these cases, immediately after division of the *medulla cervicalis*, one or two spontaneous so-called respirations, also accompanied by shivering of the whole body. This shows the extraordinary state of excitement of the spinal cord of such animals.

I obtained similar, and even more favourable conditions, when

c

very small doses of strychnine were administered to young animals, so
that the general muscle spasms were only weak, and did not continue
for a long time, and soon disappeared without the heart ceasing to
beat. After administration of strychnine, I have never seen complete
absence of muscle spasms in the animals, as Langendorff observed
with newly-born animals. In my animals, there appeared whole series
of diaphragm spasms, not only during the strychnine spasms, but also
during the interval, and especially when artificial respiration was
going on at the time the general muscle spasms approached. We
might count eight spasms per minute, and even more, if we like to
regard as full spasms all the irregular contractions of the diaphragm,
which broke sometimes the natural spasms, and also all the single
modified artificial respiratory movements which more or less came
into play. Fig. 14 and fig. 15 give such examples. During the pause
of artificial respiration, the number of the spasms of the diaphragm
was smaller, their height less, as is shown by fig. 16. It will be
easily understood that it was only necessary to touch very gently
the skin of the animal to which strychnine had been given in order
to obtain general muscle spasms, and, at the same time, produce respiratory spasms of the diaphragm. It is, therefore, not strange that
very young animals, to which strychnine has been given, behave as
Langendorff has described. Animals in this condition are more like
the reflex frog than the older examples of their own species. In the
interpretation of the appearances lies the fundamental difference
between Langendorff and myself. In beheaded animals I have never
observed *normal respiratory movements*, but always *respiratory
muscle spasms*, the same as were obtained by direct electric stimulation
of the spinal cord. The fact that older animals as well as most young
and many new-born animals, cease to breathe immediately after
division of the *medulla oblongata* below the *calamus scriptorius*, is
accounted for by Langendorff partly as an action of the vagus and
trigeminus roots, the stimulation of which, caused by the division,
has an inhibitory effect on respiration. At the same time, division of
the *medulla oblongata* causes, according to him, paralysis of the parts
of the spinal cord near the injury through shock, so that especially
the reflexes are destroyed. In consequence of the state of shock into
which the *medulla spinalis* falls by division, not only does stoppage

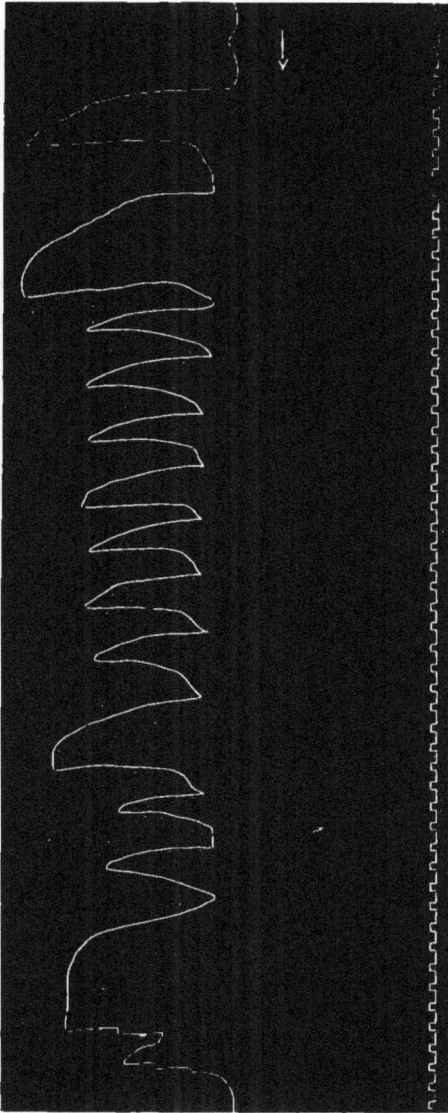

Fig. 14.—Spinal-cord respiration in a young kitten, to which strychnine had been given, during artificial respiration, after separation of the *medulla oblongata*.

Fig. 15.—Spinal-cord respiration in a young kitten, to which strychnine had been given, after separation of the *medulla oblongata*, and while artificial respiration was going on.

of respiration follow, but there is also a disposition to fatigue of the spinal respiratory centres, a property not peculiar to the latter.

This view is partly forced, and partly it does not account for the facts. At a later period I will show that the vagi nerves are not inhibitory nerves for respiration, and that, further, the fifth nerves are only subsidiary nerves of respiration, whose division does not alter the type of respiratory movement. We see, on the contrary, after division of the *medulla oblongata*, above the centre of respiration in the fourth ventricle, and of the vagi in the neck, respiratory spasms continuing

Fig. 16.—Spinal-cord respiration in a young kitten, to which strychnine had been given, after separation of the *medulla oblongata*, and after suspension of artificial respiration.

for a long time. The animal lives in these circumstances much longer than if the vagi had been left untouched. The state of shock, should it appear at all, soon passes away, and the reflexes again become normal; so it is not probable that the reasons urged by Langendorff are to be made answerable for the cessation of respiration after *sectio bulbi*. In addition, the following experiment, which I have often made, contradicts the interpretation of Langendorff. If we isolate the respiratory centre from all the centripetal nerves, whose influence comes in question with regard to the accomplishment of respiration, respiration does not stand still, but the characteristic respiratory spasms appear, which will occupy us later on. These spasms continue for a long time, and under their influence the animal becomes cold-blooded. If we divide the *medulla oblongata* from the cervical spinal cord, respiration nevertheless stops. Only one or two respiratory spasms at long intervals still appear, accompanied by rising and shivering of the whole body. On irritating the skin of the tail and extremities, one may cause one or two reflex spasms of the diaphragm. Then all is over. According to my conception, respiration stands still after *sectio bulbi*, because when the *medulla oblongata* is divided, the respiratory centre is divided at the same time. But how, then, are we to explain the respiratory spasms in the newly-born, in young animals to which strychnine has been given, and also in older animals which have become cold-blooded, especially when the spasms occur rhythmically? Not, as Langendorff says, because the spinal respiratory centres have recovered and are now stimulated by reflex action, as well as working automatically; but we must presume that, by the artificially-heightened excitability of the spinal cord, slight stimulations, let them be from the skin or the surface of the incision, suffice, by summation, to gradually cause tetanic contractions of the diaphragm, as well as of other muscles. Thus we can obtain them by means of direct electric intermittent stimulation of the spinal cord. But even if we must assume that the natural stimulation from the surface of the incision or from the skin is not intermittent, but continually active, still the rhythmic movements of the diaphragm, such as Langendorff and I have observed, can be easily explained.

It is a known and often-confirmed fact that continuous stimulations can produce rhythmic movements. It is remarkable that already at an early period, particularly in the diaphragm, rhythmic movements have been observed after death of man and of animals. Budge wrote, in the year 1842: "After the death of the animal, the diaphragm often remains in continual shivering movement for half an hour, often much longer. We see a continual swinging backwards and forwards, which sometimes, especially in older animals, continues longer than the beat of the heart." The same was observed by Valentin and Volkmann. Remak saw the movement of the diaphragm 48 hours after death, and after every sign of life in the nervous system had disappeared. This is confirmed by Brown-Séquard and Vulpian. Richet, in the year 1881, observed spontaneous movements of muscles in a dog, after death, and these lasted about 55 minutes. In the meantime the heart had been removed, and the circulation no longer existed. "These

remarkable contractions of the muscles," says Richet, "were rhythmic."
There were quiverings of the anterior extremities and of the diaphragm, which followed one another at intervals of about 4 to 10
seconds. Similar observations have been made on other muscles after
death—for instance, on the muscles of the face in the bodies of
persons who had died from cholera (M. Brandt), and in the muscles of
amputated limbs (Bennet Dowler), &c. Biedermann also found that
the muscle substance possesses the power, even by continuous action
of certain chemical stimulants, of passing into a rhythmic condition of
excitability. Well known and studied is the experiment, which consists in dipping the hinder extremities of a reflex frog into a very
dilute solution of acid (0·5—1°/$_\infty$ sulphuric acid), as has been done by
Türck, W. Baxt, and Stirling. After the frog has remained quite quiet
for a certain time in this solution, it suddenly begins, first with the
toes quite gently, then with its legs, to make a series of rhythmic
movements; when these are over, the legs may again rest tranquilly
in the acid. I have allowed the frog to register its movements

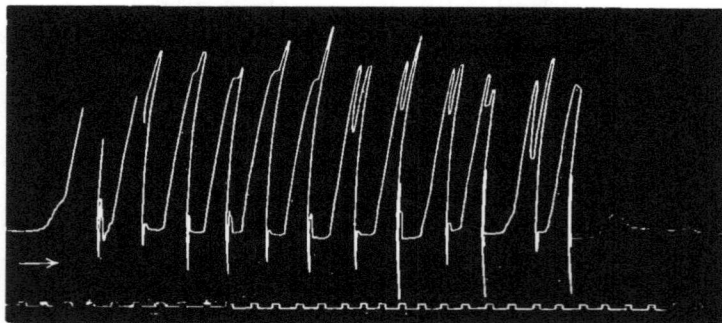

Fig 17.—Rhythmic movements of the leg of a decapitated frog which had been immersed in 0·5
per thousand solution of sulphuric acid.

by a method similar to Stirling's. Fig. 17 shows them; and every-
one will admit that they are very like the normal respiratory
movements of the rabbit. Still, no one will assert that under these
circumstances, either in the spinal cord or in the legs of the frog, there
is an active spasm centre which liberates rhythmic movements. But
who knows! For those who recognize a centre in every ganglionic
cell, this presumption may be tempting. I come, therefore, to the

following conclusions:—Experiments by means of section and stimulation have proved that in the spinal cord lie the centrifugal tracts of respiration, which convey the stimulations sent out from the respiratory centre through the ganglionic cells, in a modified way, to the peripheral nerves. *In the spinal cord there are no special centres for respiration.* The single or rhythmic so-called respiratory movements, which are liberated from the spinal cord in newly-born animals, and in animals which have been made cold-blooded, are not normal respirations, but tetanic conditions of respiratory muscles, with which, often enough, other muscles take part. All the phenomena in such cases can be explained by well-known facts. We must, otherwise, believe in the existence of centres of movement in the muscle substance, to which view Ranvier and Richet, perhaps also Lovén and Wedenskii, seem to incline. Then we would have numerous centres in the periphery of the body.

III

THE RESPIRATORY CENTRES IN THE MEDULLA OBLONGATA.

A.—POSITION OF RESPIRATORY CENTRE.   HEAD-DYSPNŒA.   MONSTERS.

That at the beginning of the spinal cord there is a part, the destruction of which immediately stops respiration and life, was already taught by Galen. Much later Lorry, without knowing the works of Galen, was led to the same conclusions by his sections of the spinal cord. "If you insert a knife or stiletto between the second and third vertebræ into the spinal cord," he says, "then you do not obtain convulsions, but the animal dies on the spot, and pulse and respiration cease for ever." Cruikshank found also that division of the spinal cord at the upper part of the neck is followed by immediate death. But Legallois was the first to point out the spot "on which respiration depended" in the *medulla oblongata*, near the origin of the vagi, so that Percy, who was present at Legallois' experiments, could declare in his report to the French Academy: "le premier mobile ou le principe de tous les mouvements respiratoires a son siége vers cet endroit de la mœlle alongée qui donne naissance aux nerfs de la huitième paire." Budge and Volkmann took away the brain from animals down to the *medulla oblongata*, namely, *cerebrum, corpora quadrigemina, pons,* and *cerebellum,* and found that this had not the slightest influence on respiratory movements, whilst a cut made in the region of the *calamus scriptorius* immediately stopped respiration. Volkmann also divided repeatedly the *medulla oblongata* along the median line without destroying respiratory movements, or even disturbing their rhythm. Flourens then attempted to fix more closely the position of the respiratory centre. At that time (1842) he laid the upper boundary immediately above, and the lower one three lines below, the origin of the vagi. The whole range in rabbits therefore consisted only of a few lines. At the same time, he guarded himself against the possibility that the origin of the vagus itself could be the respiratory centre, because the vagi in the neck might be divided and nevertheless respiration continued. Longet found, in the year 1847, that the spot pointed out by Flourens as the position of the respiratory centre did not possess in its whole thickness the property of destroying respiration, but only those intermediate bundles lying on the same level, which, rich in grey substance, cells, and blood-vessels, were very well adapted to form a special store of innervation in the centre of the *medulla oblongata.* In consequence of renewed experiments, Flourens came to the discovery of his "*nœud*," or "*point vital,*" which since has become so famous, and which he laid at the point of the **V** of the grey substance; he compared its size to that of the head of a pin, and the disturbance of it alone was sufficient to suddenly destroy life. Later, he modified this view himself in so far as he laid the position of the "knot of life" on both sides of the median line, and estimated its size at 5mm.

This last view found many opponents in addition to Longet. Brown-Séquard was already of opinion that want of the so-called *nœud vital* often neither influenced voluntary movements nor the perceptions of the mind, and even that after complete division of the *medulla oblongata* from the spinal cord, respiratory movements could be carried on. When sudden death took place, this arose from stopping of the heart in consequence of stimulation of the neighbouring parts of the *medulla oblongata,* and this could be avoided by previously cutting both vagi. The cessation of respiration was also due to the effect of stimulation. Schiff found that perforating the point of the *calamus scriptorius* had no very direct influence on respiration, and after his experiments he laid the respiratory centre in the *faisceaux intermédiaires ou latérales* (Longet). Each half of the body has, in his opinion, its own respiratory centre; both centres are separated from one another by a tolerably broad mass of grey substance. They lay, according to him, a very little behind the deep origin of the vagi, near to the lateral edge of the grey mass which forms the floor of the fourth ventricle, and they reached not so far back as the *alæ cinereæ,* the hinder part of which Schiff could remove without any immediate danger to life, while injury to the upper and outer part at once arrested respiration. So matters stood till the year 1873, when Gierke, as a result of sections, and microscopic investigations of the cut part, declared that he was unable to identify a certain group of cells in the fourth ventricle,

the destruction of which alone caused cessation of respiratory movements and the death of the animal. Very careful isolated destruction
of the deep origin of the hypoglossus, or of the *alæ cinereæ* (deep origin
of vagus), was without effect, or showed only a momentary change in
the respiration. Perforating the *alæ*, and the masses of grey substance
at their sides and above them, only stopped the activity of limited
respiratory muscle groups, and only on the injured side (diaphragm
and rib muscles). When Gierke, on the other hand, injured the longitudinal bands of fine nerve fibres which arise near the deep origin of
the vagus and hypoglossus, and branch out from these, then respiratory
movements were destroyed.   These small bundles of fibres are, according to Gierke, the main conductors of the respiratory mechanism and
obtain their stimulations from different groups of cells, to which they
are related anatomically.

Gierke was not successful in isolating the groups of cells, which
were said to contain the respiratory centre and to convey centripetal
stimulations to the cells of the motor respiratory nerves.   " These
groups of cells and longitudinal bundles," says Gierke, "are the co-
ordinate parts of a whole, which still deserves the name *respiratory
centre*, although it differs very much from that of Flourens." This view
has found many supporters and many opponents. Recently, Mislawsky
has written a paper in which he comes to results quite opposed to those
of Gierke, but Mislawsky's results have not yet been confirmed.   The
bundles of Gierke, according to him, have no connection with the
movements of respiration, which continue although these bundles be
cut.   On the other hand, the movements cease immediately when one
destroys the cell groups situated in the so-called *faisceaux intermédiaires ou laterales* of Longet, and form two centres of irregular but still
tolerably constant shape on both sides of the raphe towards the inside
of the roots of the hypoglossus and lying close to these roots.   These
centres lie between the olivary bodies and the grey substance of the
floor of the fourth ventricle, and reach from the level of the base of
the *calamus scriptorius* to its angle.   The conducting tracts of these
centres, to the origins of the respiratory-muscle nerves in the spinal
cord, lie at the level where the central canal is already closed, outside
the bundles of Gierke.   I have finally to state, that Frédericq, in opposition to Langendorff, demonstrated by experiment the existence of

a true respiratory centre in the *medulla oblongata*, in which he proved that, by direct cooling of the naked organ with ice, respiratory movements are considerably slowed, and by again heating, they increase in frequency so quickly that this effect could not be attributed to stimulation of spinal centres by the slowly spreading heat, but must have been caused by excitation of a respiratory centre in the medulla itself.

All these experiments, although differing on certain points and in many instances contradictory, have still this in common, that they assume as the sole centre for respiration a limited spot of the *medulla oblongata* in the neighbourhood of the deep origin of the vagi, the destruction of which annihilates respiration for ever, whilst isolation of other parts of the brain allows respiration to continue so long as the connection with the respiratory muscles is maintained. Opposed to this, are the opinions of those who place the chief centre of respiration at quite another part of the central nervous system, and attribute only a secondary part to the centre in the fourth ventricle. Of the spinal centres and their representative we have already said what was necessary; we will deal at a later period with the centres situated higher in the floor of the third ventricle (the centres of Christiani) and in the *corpora quadrigemina* (the centres of Martin and Booker, Christiani, and Steiner).

The experiments on the division of the *medulla oblongata* at different heights, were followed by observations of the greatest importance, in relation to the question as to whether the *medulla oblongata*, or parts of the brain situated higher, are the seat of the centre of respiration. The observations have reference to the appearance of head-dyspnœa. It was stated by Legallois, that in respiration four simultaneous movements could be distinguished: 1. gaping or yawning; 2. opening of the glottis; 3. raising the thorax; 4. contraction of the diaphragm. If we remove the *medulla oblongata*, says Legallois, all these movements stop at the same time. An incision at the level of the first cervical vertebra and simultaneous division of the vagi in the neck, stopped thoracic and abdominal respiration and also paralysed the glottis; only the gaping movements (bâillement) continued, the animals gasped for breath for some time. The same was observed by Mayers in 1815, by Bell in 1832, and by many others. Flourens observed also that when he divided the *medulla oblongata* below the origin of the vagi, the deeper the incision was, the longer the following symptoms lasted, viz.: first, lasting from 1 to 2½ minutes, a convulsive widening of the nostrils; then repeated snapping of the mouth for air; after this, the animal died. On the other hand, when he divided the *medulla oblongata* above the origin of the vagi, then the movements of the head were immediately annihilated, while those of the trunk remained intact. Volkmann also drew from his experiments the conclusion that the respiratory movements in each part of the trunk immediately cease when their connection with the respiratory centre is interrupted, while in a decapitated rabbit, especially if it be a young one, the respiratory movements of the mouth and nose continue for a long time. If a man or an animal be beheaded, says Valentin, so that the *medulla oblongata* remains attached to the head, then the jaws snap, and the trunk, on the other hand, remains to all appearance quiet. If the section of the *medulla oblongata* is situated above the roots of the vagi, then respiratory movements of the face disappear, and those of the body continue. This well-known observation can be easily demonstrated at any time. The appearance of head-dyspnœa forms therefore an infallible sign of the still existing connection between the respiratory centre on the one hand, and the facial and motor roots of the *trigeminus* on the other hand, and at the same time, the non-appearance of head-dyspnœa points to a division between them. An incision at the level of the *alœ cinereœ* is not followed by head-dyspnœa, and in these circumstances the respiratory centre must be cut off from the centres of the facial and trigeminus. It cannot, therefore, have its seat in the higher parts of the brain, whose connection with the nerves in question is untouched. On the other hand, an incision which fell between the centre of the facial and trigeminus on the one side and a respiratory centre in the third ventricle on the other would cause every movement of the nostrils at respiration to disappear. This, however, is not the case. I have made such a division, and after it have seen a lively play of the nostrils and of the muscles of the jaw.

In working out our knowledge of the seat of the respiratory centre, attention has been directed to monsters which come into the world with great defects of encephalic development and which still breathe for a considerable time. Lawrence described a case in which a child was born with deficiency of cerebrum and cerebellum and nevertheless it breathed actively. The *medulla oblongata* extended only about an inch beyond the *foramen magnum* and formed there a soft swelling which protruded at the base of the skull. All nerves from the fifth to the ninth pair were connected with this mass. On the other hand, there was no trace of feeling, and nothing seemed to point to the existence of voluntary movement. Lallemand saw a monster born without a brain, which lived for three days, emitted tolerably loud sounds and suckled. The cerebrum and cerebellum were completely absent. At the base of the skull, there was only the *medulla oblongata* and the *pons Varolii* involving the origins of the pneumogastric, trifacial, and ocular nerves.

There are many other causes recorded which prove that even the presence of the *medulla oblongata* in a rudimentary state was sufficient to support the movements of respiration.

I have undertaken no isolated extirpations in order to show exactly the anatomical seat of the respiratory centre. This is not only extremely troublesome and difficult, but gives uncertain and deceptive results; in so far as in small animals no division of such confined parts can be made without irritation, injury, or even destruction of neighbouring parts. Still, I have endeavoured to obtain a physiological conception or idea of the seat of the respiratory centre, by means of an exact analysis of the results of divisions at different levels above the respiratory centre and in the region of the same, as well as by experiments of stimulating the *medulla oblongata* directly, or by stimulating the centripetal fibres of the vagus in the neck. From all these investigations I am forced to the conclusion— *that the respiratory centre must stand in close connection with the origin or the nucleus of the vagus, and that it is very probably identical with the same.* The attentive reader of this work will no doubt come to a similar conclusion.

B.—RESPIRATION AFTER DIVISION OF THE MEDULLA OBLONGATA ABOVE THE RESPIRATORY CENTRE.   PERIODIC RESPIRATION.

We divide in a rabbit the *medulla oblongata* completely across, close below the *tubercula acustica* of the fourth ventricle, by lifting up the cerebellum in order to expose the floor of the fourth ven-

tricle; or still better, without opening the latter, by cutting straight
through the cerebellum just between the posterior and anterior lower
lips, taking care not to injure the basilar artery. The incision is
allowed to pass to the front of the *medulla oblongata*, just between
the anterior pyramids and the *pons*, running immediately under or
through the *ponticulus*. If this is done, and all bleeding has been
avoided, then respiration will go on in quite a normal condition, and
the animal will live many hours. The separation of the medulla is
made with a thin, blunt knife of a spatula form, which is quickly
stabbed through the middle of the substance. In order to make sure
that all the side parts have been cut, it is recommended, by means of
a fine, slightly-bent, blunt dissection needle, to go into the wound and
to carry the needle round to both sides towards the bones, and to let it
glide along the latter. With a little practice, one will be able, without
difficulty, to pass the knife perpendicularly to the axis of the medulla,
after the head of the rabbit has been bent a little over at a right angle
to the axis of the body, and thus, without opening the fourth
ventricle, to divide the medulla at the proper level, through the cerebellum. This method of dissecting has the advantage that no blood
can flow on to the respiratory centre; also, the latter is better protected from drying when not laid bare. In all cases I have always
convinced myself afterwards, by *post mortem* examination, of the fact
that the division was complete. I have also scorched the brain substance several times above the incision to a large extent, by means
of a galvano-cautery, in order to make certain that there was no
connection with higher parts of the brain. Besides, one soon learns
from the type of respiration if connections with the upper parts still
exist, especially after the vagi in the neck have been divided, as I
shall describe later on. Immediately after successful division, no more
reflexes can be liberated from the head. The eye of the rabbit stands
wide open, the nostrils are collapsed, neither irritation of the cornea
nor of the mucous membrane of the nose is answered—a sign that the
trigemini are out of action. The origin of the facial often comes,
quite untouched, to lie above the incision; sometimes it falls into or
under it, when the incision has been made a little too high. Respiration stops for a moment, then begins of itself. When the division is
made deeper, the pause of respiration lasts longer, and then sometimes

it requires artificial respiration to restore breathing. Once the shock is over, respiration remains quite regular. The rate of respiration after successful division of the *medulla oblongata* varies greatly in different cases. When none of the neighbouring parts are injured, it is the same as normal. I found in different animals, 82, 76, 67, 54, 52 respirations in one minute, without any dyspnœa,—a pure diaphragm respiration.

If the animals are not touched, they breathe quite regularly like machines, and generally in this way—that the respirations periodically increase and diminish in size, just as Mosso observed, and represented by diagram, as occurring in sleeping men and animals.

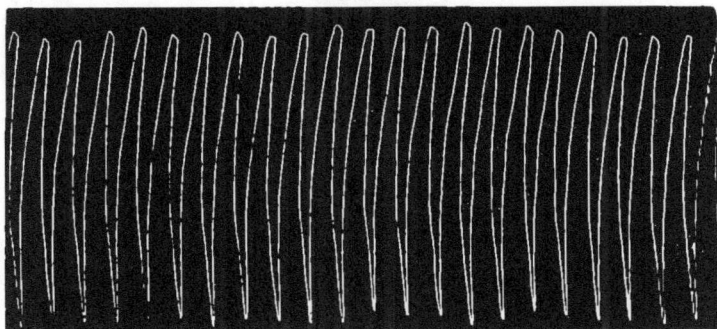

Fig. 18.—Diaphragm respiration of a rabbit whose *medulla oblongata* had been cut completely through immediately under the *tubercula acustica.*

Fig. 18 shows the respiration of a rabbit after complete transverse division of the *medulla oblongata* immediately below the *tubercula acustica.* But the further the incision has gone down (and the very smallest difference in level is in this region of the greatest influence) then the greater the difference of the respiration from the normal. At first it remains quite regular, but soon becomes slower and requires greater exertion; and, in short, it assumes the character of the respiration shown by normal animals after division of the vagi. If we cut still lower down, till we are at the extreme visible point of the *alæ cinereæ*, then respiration immediately becomes periodic. After long pauses, there follows a series of two, three, four, and sometimes even five respirations, of which the first is always the largest, while

those following gradually fall off in size.   In fig. 19 are shown periods,
each of three breaths. Quite similar periodic respiration is occasionally
seen also after division of the *medulla oblongata* at a higher level, in
cases where, through a blood-clot in the region of the *alæ cinereæ*,
pressure has been exercised on the respiratory centre.   When the clot
is removed, periodic respiration again disappears, as I have often
observed.   Further, after perfect division of the *medulla oblongata*,

Fig. 19.—Periodic (diaphragm) respiration of a rabbit after transverse division of the *medulla oblongata*
at the level of the external visible points of the *alæ cinereæ*.

respiration, which before was quite regular, becomes in time
periodic, when the centre of respiration has been exposed to the air
for any time, an occurrence which often could not be avoided.   If,
during the division, the point of the *alæ cinereæ* has been touched,
respiration becomes discontinuous; the individual breaths follow one
another at long pauses, and already bear more of the character of
respiratory spasms.   When the division is practised still lower down,
respiration is immediately destroyed, and cannot be again revived
by artificial respiration.   The most noteworthy type of respiration,
after division of the *medulla oblongata* above the respiratory centre,
is undoubtedly the periodic, and the phenomenon requires a searching
investigation.   During periodic respiration we find that the excitability of the respiratory centre has not suffered in any way.   Skin
reflexes, by nipping the tail or the paws, even by simply touching the
cutis, not only deepen the individual breaths but also increase their
number, so that the periods of respiration are lengthened.   But even
during the pauses, the skin reflexes always liberate a complete respiratory movement, so that by means of rhythmic pinching of the skin
the periodic respirations for a time almost disappear, and respiration
can be brought to its complete normal condition.   With length of time,

of course, this result becomes less marked, the reflexes, as is well known, in general becoming easily impaired. When the periodic respiration has continued for some time, then the stage of breathing-periods gradually shortens at the cost of the pauses—with every period there appear fewer breaths. Then artificial respiration must be resorted to in order to keep the animal alive. But it is very strange that immediately after artificial respiration had ceased, even when it had been in action for a long time and thoroughly applied, the periods immediately made their appearance, and were not in any way preceded by rhythmic respiration—a further sign that it can scarcely be want of excitability of the centre that causes the periodic respiration.

In order to discover if pressure on the *alæ cinereæ* was sufficient to cause periodic respiration, as appeared in the cases where a blood-clot on the *medulla oblongata* seemed to be the cause of this form of respiration, I made the experiment, without previous division of the *medulla oblongata*, of exercising pressure in the region of the *alæ cinereæ*, not only by placing a large clot of blood on the exposed part of the medulla, but also by loading the latter, in the region above the respiratory centre, with weights. Under these conditions, periodic respiration did not make its appearance. The individual breaths, under strong pressure on the *medulla*, at first changed in size. A large one was followed by a small one, the breathing became slower, but the individual respirations followed one another at equal intervals, and there was no respiratory pause. When the pressure was still increased, the respirations became smaller and smaller, and at last breathing stopped, without the appearance of a spasmodic movement of the muscles of the body. The reflexes ceased, the cornea became insensible to pain—in short, the animal appeared to be dead in consequence of a pure paralysis of the respiratory centre. But the heart had not ceased to beat, it pulsated still, although very slowly and gently; and I was successful, by immediate interruption of the pressure and energetic attempts at resuscitation by means of artificial respiration and massage, in bringing the animal back to independent respiration in a few minutes. The revived respiration was at first extremely dyspnoic, for not only did the diaphragm and thoracic muscles become active, but all the auxiliary muscles of

D

respiration as well; quietness was only gradually restored. The experiment was repeated with the same result; and although the substance of the *medulla oblongata* was greatly damaged by the pressure, the respiration which reappeared was in no way periodic, but remained quite regular. Accordingly, by means of simple pressure on the *medulla oblongata*, one can produce paralysis of the respiratory centre without concomitant spasms, but not periodic respiration. The first condition for obtaining periodic respiration, therefore, seems to be, that at least some of the normal stimuli, coming from the brain, are thrown out of action, and thus fail to influence the respiratory centre. In these circumstances, the excitability of the centre itself seems to be completely unimpaired.

Hegelmaier has also investigated closely the respiratory movements during pressure on the brain, but he never mentions the appearance of periodic respiration. That the periodic respiration of the rabbit is to be regarded as analogous with the *Cheyne-Stokes' respiration in man*, notwithstanding that the character of the periods in the latter is different, cannot be proven off-hand. I have never been successful, in the case of the rabbit, in seeing the periods running in the same way as often occurs in man: that is, in an ascending and descending scale. In the lower animals, the ascending series is always wanting—their first breath after the pause is always the deepest. Still, in man, one often sees periodic respiration with only a descending series, just as in the lower animals; and on the other hand, we cannot forget that it would be extremely difficult to make the conditions of experiment the same in men and animals. It would appear, therefore, that both in man and in the lower animals periodic respiration shows a similar character, and that the differences between the two are to be found chiefly in the different conditions of the experiment, not in the nature of the mechanism of respiration.

The information I gained in regard to periodic respiration in rabbits, namely, that it only takes place when at least a part of the higher brain tracts has ceased to act and has lost its influence on the respiratory centre, corresponds well with many observations made on healthy men and animals both on those suffering from disease in which there was periodic respiration, and where it had been caused by artificial means. Mosso has found the periodic, or, as he now calls it, *intermit-*

tent respiration, often of the Cheyne-Stokes' characteristic form, in healthy men when asleep, especially in old men and children; further, in dogs when sleeping under the influence of chloral. In the last-mentioned, neither the inhalation of oxygen nor artificial respiration could alter the periods, nor had asphyxia any influence on them. Filehne, by means of large doses of morphine, was able to cause periodic respiration in the early stages of poisoning in rabbits and dogs. A similar state of matters was observed by Langendorff after muscarine and digitaline poisoning. It may also be observed in animals in a state of hibernation (Luciani, Fano, Mosso, Langendorff, &c.). Further, Luchsinger saw, and later Langendorff and Siebert, that frogs, in which the flow of blood to the brain has been cut off, maintain a periodic respiratory rhythm, before the same became slower and irregular and finally stopped altogether (a result quite analogous to that obtained by division of the *medulla oblongata* at different heights above the respiratory centre or actually through it). When the ligature round the vessels was removed at the proper time, then the respirations again began with periods before they became regular. Martius also studied more closely the behaviour of the "salt-frog," and found that in the first stage the animals behave as if the cerebrum had been removed, and sit breathing away quietly. In the second stage, periodicity of the respiration sets in, while in the third stage only single and irregular respirations take place, which are often caused by stimulation at the periphery, and the animal soon dies. The analogy of these observations with those I have obtained by sectional experiments becomes more complete when we consider the experiments of Sigmund Mayer, who proved that in the execution of the Kussmaul-Tenner experiment the upper parts of the brain are first thrown out of action and then finally the respiratory centre; while with renewed blood current it is the opposite: first the respiratory centre becomes active, and then by degrees the other parts of the brain—*i.e.* the later the further up they lie. That it was only necessary to partially throw out of action the upper tracts of the brain was proved to me by the observation of a patient in the clinique of Prof. Lichtheim in Bern. This patient showed the modification of the Cheyne-Stokes' respiration which we observe in animals, namely, the absence of the ascending series, while the descending series and pause (although the latter was short) were quite plainly marked. The patient suffered from hemiplegia, probably the result of a softening caused by arteriosclerosis. She was quite conscious, and answered all questions quickly and correctly; she was sensible to pain, and was also able voluntarily to change the time of the respiratory movements; but when she was left to herself, the periodic respiration immediately appeared again, and both sides of the thorax moved at the same time. Therefore, in this case, one-sided deficiency of the upper brain tracts was sufficient to maintain the Cheyne-Stokes' respiration as long as the attention of the patient was not directed to its control, but even then the periodic type did not completely disappear.

There are also a great many experiments proving that the excitability of the respiratory centre does not suffer much during periodic respiration. In cases in which periodic respiration was brought on by cutting off the flow of blood to the brain, Langendorff and Siebert were able to liberate a whole group of respirations during the interval between two periods, by stimulating the skin (reflex action).

Further, Valentin, who made a long series of experiments regarding the respiration of marmots, reports that, during the deep sleep of these animals, while respiration is very slow and irregular, by nipping the paws or tail, at every pinch the animal gives a deep respiration. I can confirm this from personal observation. Also, the discoveries made by Luciani, Fano and others on patients, showing the Cheyne-Stokes' respiration phenomenon, are in favour of continued excitability of the respiratory centre during the respiratory pauses. These experimenters were successful, by addressing the patient, in interrupting the respiratory pauses, that is, in liberating respirations during the pause. But if one has obtained periodic respiration in rabbits by extirpation of the higher brain tracts, and then cut the vagi in the neck, periodic respiration disappears immediately and gives place to irregular respiratory spasms; and one is no longer successful, by any means, in causing periodic respiration. This discovery had also been made by Traube. On the other hand, Filehne maintains that he saw periodic respiration continue after division of the vagi; his curves are truly not convincing, and my observations speak thoroughly against his views. That the respiratory centre, even after division of the vagi, retains its complete excitability will be discussed afterwards when skin reflexes and stimulation of the central ends of the vagus in the neck are under consideration. As a second condition for the appearance of periodic respiration, it therefore seems to be necessary that the peripheral branches of the vagi should be in connection with the respiratory centre. I have intentionally kept myself free from every theory which could explain the manner in which the rhythm of the Cheyne-Stokes' respiratory phenomenon is accomplished. So long as one does not know exactly all the conditions under which this type of respiration appears, all explanations, let them be ever so learned, are worse than useless. I have been content to indicate some of the conditions which must necessarily exist, and at the same time I have endeavoured to show, from my own and other investigators' experiments, that in this the respiratory centre itself does not play the chief part. Because when periodic respiration (no matter how distinctly it has shown its character before) stops immediately after division of the vagi, and when at the same time it is proven that the respiratory centre, during the respiratory period as well as during the respiratory pause, can be equally stimulated by equally strong shocks, then the reason for the Cheyne-Stokes' respiration cannot be found in a variation of excitability of the centre, as is thought by Luciani and Fano, but must be sought outside the respiratory centre. Traube also attributed the occurrence of periodic respiration to the altered excitability of the centre as well as to the action of the vagi. I am inclined to take an opposite view of the question. While the excitability of the centre is unaltered, there are acting on the centre, either from the vagi or from the higher brain tracts, stimulations which are no longer continuous or which perhaps only change in intensity. If the upper brain tracts are completely cut off (as after separation of *medulla oblongata* from the brain) then periodic respiration is connected only with the presence of the vagi and disappears when they cease to act. Even stimulation of the cut central fibres of the vagus in the neck by intermittent shocks can only provoke regular, but never periodic respiration. Still it may be possible even in this way, by gradually increasing and decreasing stimulations or by means of properly graduated rhythmic stimulation, to cause periodic respiration.

C.—AUTOMATIC ACTION OF THE RESPIRATORY CENTRE.

After having thoroughly considered the effects following different sections of the *medulla oblongata* above the respiratory centre and through the same, there arises the important question, whether the centre, after being cut off from all centripetal nerves, is still capable of liberating respiratory movements; that is, can it be automatically active? Up till now this problem has only been incompletely explained. Flourens divided brain and vagi and saw respiration continue, although with difficulty. Volkmann removed the brain, the vagi, and the lungs, taking care not to injure the phrenic nerves, in a kitten (also in dogs), and saw respiration continue for forty minutes. As Rach then maintained that division of the posterior nerve-roots of the spinal cord in the neck, with or without previous section of the vagi, was sufficient to stop respiration immediately, Rosenthal again commenced sectional experiments, and attempted as far as possible to isolate the *medulla oblongata* from all centripetal nerves, by dividing the posterior roots of the spinal cervical nerves and the whole spinal cord at the level of the first thoracic vertebra, and also by cutting the vagi and separating the *medulla oblongata* in the region of the *corpora quadrigemina*. Still the animals breathed regularly, says Rosenthal, and at the following rates:—after division of the posterior roots, 14 respirations in 15 seconds; after division of the spinal cord, 22 respirations in 15 seconds; after division of the vagi, 11; after a quarter of an hour, only 6 in 15 seconds. After division of the *medulla oblongata* the rate of respiration was not altered. The last statement shows, as we shall soon see, that the division of the *medulla oblongata* was either too high or must have been incomplete.

I have often repeated Rosenthal's experiments. The *medulla oblongata* was cut across immediately above the respiratory centre, also the spinal cord at the height of the last cervical vertebra, and then the vagi and superior laryngeal, as well as glossopharyngeal, were ligatured or cut. The animals were then no longer capable of regular respiration, as Rosenthal maintained; but there appeared inspiratory and expiratory respiration spasms, in which were involved the muscles of the thorax—one, two, to three spasms in a minute, similar to the respiratory spasms as they appear after division of the *medulla oblongata* and the vagi alone, only still more dyspnoic. The reflexes, which could only be caused by pinching of the fore-paws and the skin in the region of the neck, liberated inspirations and expirations, but the latter only seldom. If the brachial plexus was then divided, so that the *medulla oblongata* only stood in connection with the few centripetal nerves which come from the neck and those which run into the spinal cord with the phrenic nerves, and which are not capable of replacing the non-activity of the vagi, or upper brain tracts, then the respiratory spasms remained unaltered. That such animals generally died half an hour after the operation is not only explained by the character of the respiratory spasms, but is also partly owing to the effects of the division of the spinal cord, an operation which rabbits do not very well sustain. I therefore come to the conclusion that the respiratory centre in the fourth ventricle can be automatically active, but the automatic mechanism manifests itself only in inspiratory and expiratory spasms.

D.—DIRECT STIMULATION OF THE MEDULLA OBLONGATA. INSPIRATION AND EXPIRATION CENTRE.

To secure direct stimulation of the *medulla oblongata* above the respiratory centre, two fine English needles were used, which were isolated and passed through a cork, and inserted immediately below the incision, which divided the medulla in the region of the *tubercula acustica*. The medulla was then stimulated by means of the arrangement already described, with single shocks as well as by an intermittent current. In order to make single shocks effective strong currents were required—currents of 500 to 1000 Unit. when a Du Bois-Reymond sledge inductorium, worked by two Daniell cells, was employed. If we now stimulated during normal respiration, then we obtained, with single opening shocks (closing shocks had no effect), at the beginning of the inspiration, a lengthening of the inspiration. At the top of the inspiration the inspiration was often interrupted, and an expiration was liberated. When we stimulated during the respiratory pause, then an inspiration immediately took place. Sometimes we were successful in inserting artificial respirations between the normal respiratory curves, by means of an opening induction shock, which were quite similar to the normal respirations. As a general rule, when we stimulated during expiration an inspiration appeared, and on stimulating during inspiration often an expiration appeared. When the stimulation took place during periodic respiration, then a single shock liberated a simple respiration, which, for depth and duration, was equal to a normal one, and that at the end of the period as well as during the pauses, so that respiration could be

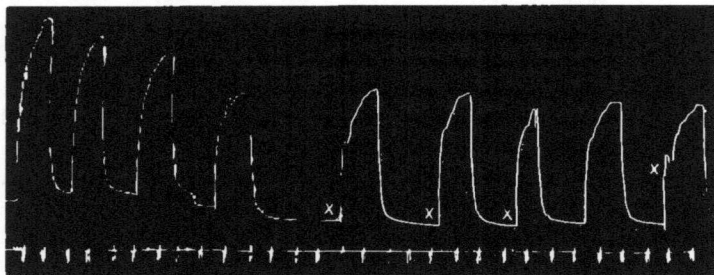

Fig. 20.—Direct stimulation of the *medulla oblongata*, above the respiratory centre, by means of single induction shocks (300 Unit. 1 D.) during a pause between two attacks of periodic respiration. The *x* denotes moment of stimulation.

maintained in a rhythmic manner for a considerable time (fig. 20). When the animal was made apnoic by means of artificial respiration, then during the apnœa even the strongest single shocks could not liberate respirations. In cases where, after division of the *medulla oblongata* with injury to the *alæ cinereæ*, the respiration was deficient and dyspnoic, single shocks could not liberate a respiration during the pause; and only when the single shocks acted at an interval of some seconds did one see that after a long series of such shocks the respiratory periods became lengthened by supernumerary respirations. Further, it was found that there was an analogy between the observations of Bowditch and Kronecker concerning the change of excitability of the electrically-stimulated ventricle of the frog's heart, and those made by me, with the following results:—The stimulations, which followed at intervals of some seconds, gained in effect, so that shocks which before were just sufficient, could now sometimes be considerably weakened before they became completely ineffective;

but after that, they had to be again brought back to their previous
high strength in order to have effect. If the *medulla oblongata*
be stimulated by means of intermittent shocks, at an interval of
from $\frac{1}{12}$ to $\frac{1}{30}$ of a second, either continuously or rhythmically,
then shocks of far less intensity are sufficient (70 to 200 Unit.),
during normal respiration, to quicken the same, and, during periodic
respiration, to lengthen the periods, by increasing the number of
respirations during the period. In these circumstances, respiration
may become quite regular, so that the pauses entirely cease. In
this way, by rhythmic electric stimulation, it is possible to maintain

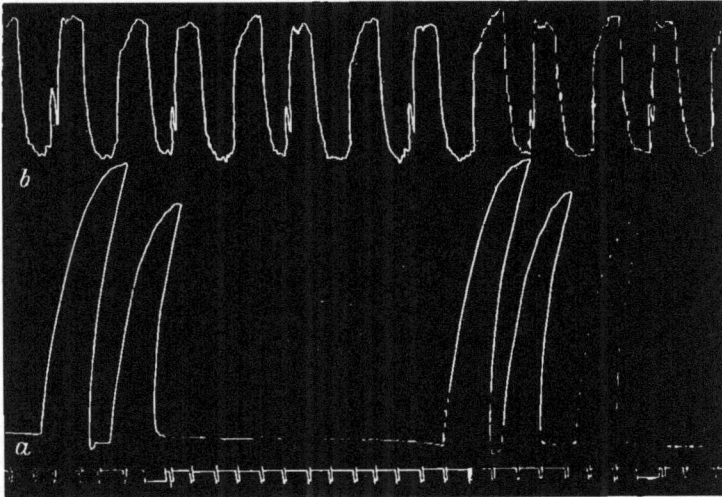

Fig. 21.—Direct stimulation of *medulla oblongata* by intermittent currents ($\frac{1}{12}$" interv. 175 Unit.) during
periodic respiration. *a*. Periodic respiration; *b*. respiration during the stimulation.

artificial respiration by reflex action (fig. 21). Thus the diaphragm
generally becomes increased in tonus, and therefore respirations are
accomplished while the diaphragm descends more in an inspiratory
position, and single respirations become lower. Generally the tendency
to inspiratory movements prevails, so that sometimes one respiration
is followed by a second before the diaphragm has time to get into a
position of rest.

If one stimulates the *medulla oblongata* with strong intermittent currents (600 to 2000 Unit.) for a considerable time, then follows an intense inspiratory spasm (tetanus) of the diaphragm, which disappears very slowly, and which is often followed by a strong active expiration, and even by a second inspiratory spasm. During the spasm of the diaphragm, thoracic respirations also often appear, and can be recognized as teeth on the inspiratory curve. In such cases, one may observe the action of the heart becoming slower, and then stopping altogether. At the same time, opisthotonus and clonic spasms of the muscles of the body appear. Such a spasm sometimes is followed by a series of independent normal-looking respirations after the stimulation has ceased, even when originally the animal had automatically breathed intermittently or defectively. During these respirations, the diaphragm gradually goes back into its original position of rest. If, after the *medulla oblongata* had been successfully divided above the respiratory centre, the vagi were cut in the neck, so that respiratory spasms appeared, and if then the *medulla* was stimulated rhythmically or continuously with intermittent shocks, one was always successful in modifying the respiratory spasms and in liberating rhythmic respirations. This was done especially at the cost of expiration; but for all that, one was not able, by means of electricity, to make the respirations as frequent as before division of the vagus. The respirations were certainly quite regular, and followed one another without pauses, but all the same the inspirations remained spasmodic, although not in comparison with those before stimulation. In fig. 22 the respirations obtained by rhythmic electrical stimulation of the *medulla oblongata* are figured in the curve of a previous respiratory spasm, so that one can see the difference between the two. As soon as stimulation was interrupted, the inspiratory spasms became still longer, followed by shorter and longer pauses. Although, on the whole, direct stimulation of the *medulla oblongata* brought forth more inspiratory movements, yet I often saw such plain and active expirations appear that I could not doubt the existence of an expiratory centre of respiration as well as an inspiratory one. But the expiratory centre of respiration is not only, as is evident from the preceding, more difficult to stimulate than the inspiratory, but it only comes into action under exceptional circumstances, and never takes part in ordinary respiration. Likewise, as in the rabbit, generally only the diaphragm, and then in exceptional cases (as when respiration is impeded) the muscles of the thorax, and lastly the whole of the subsidiary muscles, become innervated in their order by the respiratory centre; so in the same way, the expiratory centre goes into action later, and only under exceptional circumstances—as, for instance, in

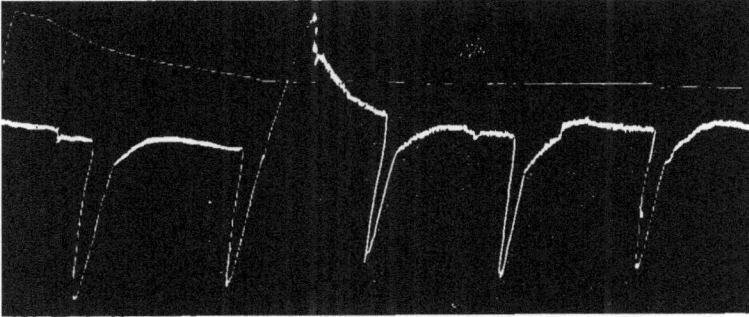

Fig. 22.—Diaphragm respiration of the rabbit, on direct stimulation of the *medulla oblongata* by means of rhythmic electric irritation, after the *medulla oblongata*, above the respiratory centre and the vagi in the neck, had been divided. The respiratory movements, liberated by electricity, have been figured in the curve of a preceding respiratory spasm.

some physiological and pathological movements, such as coughing, sneezing, eructating, vomiting, as well as in certain kinds of dyspnœa. But to refer normal rhythmic respiration to the alternate stimulation of the inspiratory and expiratory centre is impossible, for this reason, that expiration, as a rule, is a passive act. It would be as absurd to assume that the inspiratory and expiratory effects obtained by direct stimulation of the respiratory centre were produced by the electric shocks alternately causing the centre of inspiration and expiration to come into play.

The most important fact gained by direct stimulation of the respiratory centre is that the centre can only be electrically stimulated in the same way as if the stimuli came from the centripetal nerves, and therefore we stimulate only the sensory ganglionic part of the centre with intermittent currents. To bring the motor-ganglion part in the centre to direct rhythmic action is attended with just as little success as in the respiratory tracts situated in the spinal cord, where we obtained no respirations but only respiratory spasms. Gathering together the results obtained in this section we arrive at the following conclusions:—In the *medulla oblongata* lie the centres of respiration in close connection with the vagi roots—a centre of inspiration, and a centre of expiration which it is more difficult to excite to action. The centres act automatically as well as by reflex action. Head-dyspnœa speaks very emphatically against the existence of respiratory centres situated higher in the brain. For the accomplishment of periodic respiration, it is necessary that at least a part of the upper brain reflexes should be thrown out of action, while the vagi still act, although imperfectly. In periodic respiration the respiratory centre is fully excitable, and can be excited during the time of the periods, as well as during the time of the pauses, by equally strong stimulations.

IV

THE CENTRIPETAL COURSES OF RESPIRATION, THEIR FUNCTION AND NORMAL TONUS.

A.—THEORIES OF RESPIRATION.

To give even a short abstract of all the different treatises on this subject which have been published, and of all the conclusions writers came to with respect to the liberation of respiratory movements, one could not only write a large volume, but in the end one would have to admit with sorrow that the results were so contradictory and the theories so different, that it had not repaid the time and trouble. I therefore refer to the bibliography at the end of this work, where the reader will find a list of the principal treatises.

The first thing naturally to be studied was the effect of the division of both vagi in the neck. With Petit and Haller, a large number of learned men began the study of these phenomena, and already at that early time the controversy arose, why animals after division of the vagi sooner or later die? This discussion went on until Traube in his classic investigations "regarding the causes of the changes which the lungs sustain after division of the vagi," gave proof that it was not the direct action of the vagi on the centre of respiration, but secondary changes in the lungs that caused death. At the same time he showed that not only Dupuis and Burdach were wrong in believing that paralysis of the lungs was the cause of death after vagotomia, because the interruption of the conversion of venous blood into arterial destroyed the chemical side of the respiratory process; but also that Brachet was mistaken in stating, that after division of the vagi the necessity for respiration ceases, against which view Volkmann had already entered a protest. This view of Traube's, however, although it was given in a very convincing manner, did not remain without opponents; for instance, Eichhorst was of opinion that in birds, and under certain circumstances also in mammals, death was not caused by vagus-pneumonia, but by the appearance of paralysis of the heart. After the section-experiments followed others, who studied the effect of mechanical and electrical stimulation of the vagi. Krimer (1820) appears to have been the first who proved the effect of galvanism on the central ends of the divided vagi, and found that thereby respiration became quicker. Then came Marshall Hall, who showed that by stimulating the vagi in the neck of an ass with a forceps, an inspiration was produced, which was always followed by a sort of swallowing-down movement. Budge, and especially Traube, and many other experimenters followed, whose names and merits are honoured by mention in Rosenthal's famous work. By the facts gained by division and stimulation of the vagi, theories regarding respiration were advanced. This was quite natural. Haller taught, "that the exertion of the soul to maintain life caused the alternate movements of inspiration and expiration;" Whytt sought the cause of respiration in an instinctive action of the nervous system; Buffon attributed the movements to the stimulations caused by the inspired air; Darwin saw its cause only in the rush of blood to the lungs; and Borelli and Manzini found the explanation of rhythm in respiration in an antagonism between the upper and lower lung cells. This resumé is historically interesting, and shows the state of science at various periods. Then, Martin and Boerhaave were of opinion, the first, that the phrenic nerve is pressed upon by the expanded lungs in inspiration, is thus made inactive, and the diaphragm consequently relaxes; the other, that in inspiration no blood flows from the lungs into the arterial system, and the brain from want of blood yields up its activity to the muscles of inspiration. Roose, in his anthropological letters of 1803, advanced the opinion that during expiration the brain was filled with more blood, and therefore, being more sensitive, stimulated the muscles to inspiration, and that during inspiration it received less blood and therefore the action of these muscles ceased and expiration followed. Again, in the year 1814, Bartels writes, that in inspiration the brain receives more arterial blood, and, thereby excited, stimulates the vagi, and so the expiratory movement of the lungs takes place, and that then, during the latter, the brain is again put out of action by the gathering venous blood and for the moment is paralysed, and then the spinal cord works freely and brings into action the movements of inspiration. But already, in the year 1823, Rolando taught that venous blood in the lungs causes a stimulation of the vagus, as a sensory nerve; this stimulation passes along the *medulla oblongata*, from which the muscles of the thorax are compelled to make the movement of inspiration. Then Johannes Müller came to the conclusion that it was arterial blood which stimulated the *medulla oblongata* to discharge into the respiratory nerves; while Arnold assumed that the feeling of the necessity to respire depended partly upon the action on the mucous membrane of the air which got unfit for respiration, and partly upon the gathering of blood in the vessels of the lung, and that this feeling was conveyed through the vagus to the brain. So even at that time the controversy had begun as to whether oxygen or carbonic acid was the stimulus for respiratory movements.

Respiration, says Marshall Hall, is a mixed mechanism, only partly dependent on the activity of the brain or the will. The partial dependency of the excito-motor power shows itself after division of the vagi. But neither the brain nor the pneumogastric nerves are necessary for the acts of respiration, as these continue when either or both of these organs has been removed. With the removal of both, inspirations cease. Inspiration can, in fact, be an act of the will even when the vagi are divided, or it can be a reflex action, caused by means of the pneumogastric nerve after removal of the brain. Thus it is evident that the *medulla oblongata* is not the "*primum mobile*" of respiration, but it is the vagus which acts as a stimulator of respiration. It is essential and necessary, even when the power of the will has been removed with its organ, the brain. The acts of inspiration are also acts of the excito-motor system, or, properly speaking, of the spinal-cord system. Ordinary inspiration is stimulated by means of the pneumogastric nerves, but controlled and governed by the will. The pneumogastric nerve is not, however, the only excitor nerve of inspiration. Inspiration is stimulated by the fifth pair, as also by the spinal nerves (in such actions as sprinkling the face with cold water, stepping into a cold bath, the first respiration of a new-born child, as well as the first act of expulsion of fæces and urine by the child). Several facts confirm the idea that the carbonic acid in the air-cells of the lungs and in contact with the fibres of the pneumogastric nerve is the stimulating cause of inspiration. The acts of inspiration, as already stated, are excited acts, and are set into action by different excitor nerves: 1. by the trifacial; 2. by the pneumogastric; 3. by the spinal nerves. The *medulla oblongata* must be regarded as the organ which unites the different muscles into one system, and the different nerves which are included in the respiratory system of Sir Charles Bell are the proper motor nerves of respiration.   Thus wrote Marshall Hall as early as 1839.   Have we since then gained much new information?   Then follow A. W. Volkmann's numerous experiments (1841), which led him to the following conclusions:—The sensory nerve-fibres convey impressions to the *medulla oblongata* from all organs of the body. The lungs, particularly the nerves of the lungs, have no greater influence over the accomplishment of respiratory movements than the nerves of other organs.   The cause of stimulation is carbonic acid, not that in the respiratory passages, but that in the blood throughout the body; the point from which stimulation issues is every part of the body; each sensory nerve thus stimulates, conveying influences up to the *medulla oblongata,* and thus stimulations do not pass along the vagus alone. Respiratory movement is the result of the necessity to respire expressed by the whole body, and it arises from the energy with which the parenchymatous exchanges of gas go on everywhere.   The centripetal nerve-fibres stimulate, through the *medulla oblongata,* the nerves which supply the respiratory muscles.   In the organs and tissues there always exists a necessity for respiration, and thus respiratory movements are excited without concurrence of the will or consciousness. If the movement of respiration be stopped a little longer, then the peculiar feeling of dyspnœa sets in, which has its seat in the lungs, and without doubt is transmitted through the vagus and is the result of gaseous exchanges, in the substance of the nerves themselves, being hindered on account of the large quantity of carbonic acid contained in the air in the lungs, &c.   According to the above, the difference between the interpretation of Marshall Hall and that of Volkmann is on the whole not so very great.   Both accept carbonic acid gas as the stimulant which irritates the centripetal nerves, and which, on their part, liberate reflex respiratory movements (although not reflex in the present sense of the word); but while Marshall Hall makes the carbonic acid in the lungs, and through that the vagi, mainly answerable, Volkmann extends the law to all centripetal tracts. Valentin (1848) differed from both; he was of opinion that respiration did not originate in the lungs and was not caused by a reflex action.   Respiration therefore could not be liberated through stimulations caused by the inspired air on the mucous membrane of the lungs or on the nerve endings of the vagi. The cause, according to Valentin, lies in the *medulla oblongata* itself.—" The deficiency of arterial blood, or too small a quantity of suitable blood, he says, first alters the nature of the respiratory movements and at last stops them altogether. But if the *medulla oblongata* does not act of itself, then an adequate stimulation is capable of renewing its influence at least for a moment. Certainly Volkmann and Vierordt were wrong when they assumed that the influence of the venous blood acted as a stimulant from every point of the body through the nerves to the *medulla oblongata*. This took place only when that part of the central nervous system was immediately subjected to irregular conditions, so that suffocation was thus caused and not by irritation of peripheral nerves." Valentin means, therefore, that the stimulating cause of respiration is not the carbonic acid, but the deficiency of oxygen in the blood. This deficiency of oxygen directly stimulates the *medulla oblongata* to action. This view we find taken up later by Rosenthal. L. Traube, who now followed, was inclined more to the views of Marshall Hall and Volkmann, advancing the opinion that the dypnœic appearances in mammals depend not on the diminished supply of oxygen, but on the diminished liberation of carbonic acid gas which is continually forming in the body; that the carbonic acid gas is the agent which, either by direct or indirect stimulation of the respiratory nerve centre in the *medulla oblongata*, produces inspiration and expiration; and further, he held that the dyspnœic phenomena caused by the accumulation of carbonic acid in the body became the more energetic the greater the quantity of oxygen the blood contains at the same time. With Traube, again, it is therefore the accumulation of carbonic acid which becomes the stimulating cause of respiration, and he leaves it undecided whether, directly or indirectly, the *medulla oblongata* is forced by the same to the liberation of respirations.

In the year 1862, appeared Rosenthal's work, which has already been mentioned, in which he seeks to prove, with reference to the theory of respiration, that, as was maintained by Valentin, the action of the blood on the *medulla oblongata* is direct, and that Volkmann's view on the whole is wrong. It seems scarcely conceivable, says he, that the quantity of gas in the blood could primarily be altered by division of the vagi, nor could the assumption be regarded as admissible that this altered quantity of gas should act so differently from the general rule on the respiratory nerve centre in cases of difficulty of respiration (namely, by diminishing the frequency of the respirations). The vagi have at first and directly, says Rosenthal, nothing to do with the degree of activity of the *medulla oblongata.* This degree of activity appears rather to be regulated only by the quantity of oxygen in the blood; the vagi, on the contrary, are simply connected with the mechanism which regulates the rhythm. As a main proof that the irritation of the respiratory centre is caused by deficiency of oxygen, Rosenthal brings forward observations made during apnœa. The respiratory movements are caused by the irritation of the blood on the respiratory nerve centre. The transference of these stimulations to the respective nerves and muscles meets resistance, by which constant stimulation is converted into rhythmic action. This resistance is diminished by the action of the vagi, and is increased by the action of the superior laryngeal branches of these nerves. The degree of activity of the central organ is dependent on the quantity of oxygen in the blood, and the splitting up of this activity into single respirations is dependent on the activity of the nerves. The respiratory movements are not to be designated automatic, in the sense that the ganglia of the "*nœud vital*" must be active of themselves, in consequence of an inherent mechanism; on the contrary, these ganglia receive first from the blood the stimulation to activity. This theory of Rosenthal's, which he defended and enlarged in a large number of papers, found many opponents, as we shall discuss more fully further on, when the vagi are under special consideration. The question whether deficiency of oxygen or excess of carbonic acid was the irritating cause of respiration has since then often occupied science, and has at last so far been decided by Pflüger and Dohmen that both may be active. Bernstein, the last of this group of workers, made the distinction as to the action of both gases more precise, by stating that blood poor in oxygen preferably stimulated inspirations, and blood rich in carbonic acid, on the other hand, excited expiration. He is further of opinion that the reaction of the respiratory centre to blood poor in oxygen is a regulation of a useful nature, which fulfils the purpose of bringing oxygen in sufficient quantity to the blood. But the separation of carbonic acid is just as important to the organism, because an accumulation of this gas is injurious to life. Therefore the action of carbonic acid on the expiratory centre by which a considerable quantity of carbonic acid is separated from the lungs, is also a regulation of a most useful kind. Unfortunately, the convenient theory of Bernstein is not correct, as Gad has recently proved.

The consideration of the rhythm of the respiratory movements and the results of experiment showing that expiration during obstruction of respiration is accompanied by contraction of special groups of muscles, as well as the observations that by stimulation of certain nerves, expiratory movements took place so as to cause cessation of respiration, led, at an early period, to the initiation of two opposed centres for respiration—an inspiratory and expiratory centre. Budge seems to have been the first to distinguish between the two centres. He suggests: 1. A vagus centre for the movements of expiration (because he obtained expiratory effects from stimulation of the vagus); and 2. A centre of inspiration—the *"nœud vital"* of Flourens. Others soon followed, such as Hering and Breuer, and Lockenberg; more recently, Kronecker and Marckwald, Langendorff, Christiani, and others.

In opposition to the chemical theory, according to which *the blood* was the direct or indirect cause of respiration, theories were brought forward (also supported by experiments) which maintained a more mechanical and purely nervous conception of respiration. The chief representatives of this theory were Schiff, Von Wittich, and others, but especially Hering and Breuer, who found in their experiments that inflation of the lungs (with uninjured vagi) makes inspiration more difficult, and favours expiration; whilst sucking the air out of the lung, on the other hand, favours inspiration, and makes expiration difficult. They concluded that inspiratory movements of the lung called forth expiratory, and the expiratory brought on the inspiratory. This automatism (or self-regulation) accordingly was the reason of the rhythmic respiratory movements.

The Hering-Breuer theory, which I shall criticise at another place, has found many opponents and many friends, and a series of works by Lockenberg, Guttmann, and by Gad confirm the results of the Hering-Breuer experiments. More recently, Gad has brought forward a theory of his own for explaining respiration, based on the view that the vagus nerve has principally an inhibitory action on respiration. Gad is of opinion that inhibitory nerves act "favourably on the normal course of the nutritive processes. The changes in ganglia only take the form of functional activity when some derangement has taken place in the normal course of the nutritive processes. The rhythmic discharges of the central organs may be explained further by assuming that, on the whole, the conditions for disturbance of the normal course of the nutritive processes, and the removal of these conditions, do not coincide. The respiratory inhibitory fibres of the vagus are stimulated first by inspiration itself, not before; they therefore possess no tonus. This periodic vagus stimulation interferes with the disturbance of the normal course of the nutritive processes in the centre, but it disappears as soon as this result has been obtained." Christiani is of opinion that the whole process of respiration is caused by stimulation of the respiratory centre, which acts in regular periods, but is not excited by reflex action. There are three centripetal ways by which the stimulations for inspiration approach the complex ganglia which form the respiratory centre. These are, namely: (1) Through the sense organs—eye and ear; (2) through the sensory nerves of the skin; (3) through certain vagus fibres; while inhibition and active expiration takes place through the remaining vagus fibres, through the trigeminus, and through certain fibres of the other sensory nerves along which impressions pass that cause pain. Unfortunately, Christiani does not tell us where the periodic stimulations of the respiratory centre which are *not* caused by reflex action originate. There are, therefore, in this mechanical theory two conceptions—the direct and the indirect (reflex) stimulation of the respiratory centre.

When we sum up then the views of the experimentalists, they may be divided into two great groups: (1.) The one adheres to the chemical theory, and considers the irritation of the blood as the stimulus of respiration; and this some consider to be caused by deficiency of oxygen, others by excess of carbonic acid, whilst a third division thinks it is due to these two factors in common. (2.) The second group makes mechanical or purely nervous arrangements answerable for the stimulation and liberation of respiratory movements. In both groups we find defenders of the direct and indirect stimulation of the respiratory centre.

B.—EXTIRPATION OF THE DIFFERENT CENTRIPETAL NERVE TRACTS ACTING ON THE RESPIRATORY CENTRE.

It is generally and fully acknowledged that a whole series of centripetal nerves act on the respiratory centre, and are capable of altering the respiratory movements. We must then first of all prove what is the function of these nerves during normal respiration, in order to ascertain whether, in the accomplishment of respiratory movements, they are constantly active and necessary, or whether they come into action only under exceptional circumstances, and as auxiliary nerves. If one divides all the centripetal nerves coming into question in respiration into single groups, then there can be distinguished: (1.) The upper tracts—that is, the higher nerves of sense, the *trigeminus*, and those sensory nerves which come from the brain into connection with the respiratory centre. (2.) Those centripetal nerves which enter directly into the *medulla oblongata* and communicate with the respiratory centre—the vagus and glossopharyngeus. (3.) The inferior tracts, especially all ordinary sensory nerves which through the spinal cord act on the *medulla oblongata,* and the splanchnic nerves. What happens when we extirpate these individual tracts?

a) *Extirpation of the vagi and glossopharyngei after separation of medulla oblongata above the respiratory centre.*

In the foregoing section, we saw that when we separated the upper tracts by means of an incision which passes through the *medulla oblongata* beneath the *tubercula acustica,* respiration remains the same as under normal conditions. If you then divide both the vagi in the neck, or ligature them, or cool them suddenly by the method of Gad, respiration changes all at once in a very remarkable manner. The diaphragm immediately after extirpation of the vagi passes into a long inspiratory spasm—gradually, not suddenly—as if an inhibition was removed; and it always commences at the phase to

which the respiration was confined, according as extirpation was performed during inspiration or during expiration. Thus the diaphragm
in this spasm sometimes goes considerably beyond the ordinary height
of an inspiration, and then falls back suddenly into the position of
expiration. It is often passive, but often it is supported by considerable contraction of the muscles of the abdomen. This occurs
when the preceding spasm was very strong and long, and was immediately and without pause followed by a second shorter inspiratory
spasm. Sometimes, however, it is quite the opposite, the first spasm
being the shorter, the second the longer. Then long and short
inspiratory spasms alternate with short passive or active expirations.
At the beginning, there are no respiratory pauses present, but these
gradually appear, having a shorter or longer duration. I have often
observed inspiratory spasms of 1·75 minutes' duration, and longer.
During these, the animal may become so dyspnœic that along with
the contractions of the diaphragm, thoracic respiration also takes place.
These thoracic respirations are marked on the diaphragm curve as
small or larger roundish elevations in the course of the line of inspiration. While the thorax is widened, the rib movements of the diaphragm
become still further extended, and the diaphragm therefore descends
still more in the sense of inspiration. The respiratory pauses, at first
short, now gradually increase in size at the expense of inspiration,
and, at last, at the beginning of each expiration and inspiration, there
appears a thoracic respiration. The animal dies in a short time (2
to 3 hours after the operation) if artificial respiration is not soon
employed at intervals. Fig. 23 shows the beginning and end of the
respiratory spasms after the *medulla oblongata* had been transversely
divided above the respiratory centre, and the vagi in the neck had
been cooled. The duration of the spasm was 1·75 minutes. On the
under curve of plate I., one can see similar respiratory spasms after
division of the vagi in the neck; at the same time, the thoracic respirations are marked on the line of inspiration. When the incision
through the *medulla oblongata* was so low down that, before division
of the vagi, periodic respiration already existed, then section of vagi
altered the type of respiration in the same way as before: the periods
disappeared at once, and the characteristic respiratory spasms took
place. When respiration before division of the vagi had already been

interrupted (as happens after injury to the deep root or nucleus of the vagus), then after division of the vagus in the neck, there appeared very long pauses; and after these, either one, or, following close upon one another, two quick and deep (spasmodic) respirations, generally with active expiration, if death did not immediately take place.

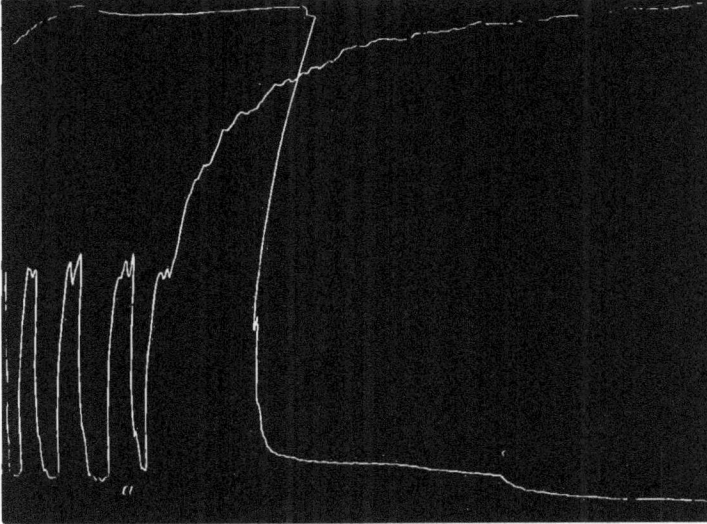

Fig. 23.—Diaphragm spasm after separation of *medulla oblongata* above the respiratory centre and cooling of the vagi in the neck. Nerve cooled to −5° C. Duration of the diaphragm spasm (of which here only the beginning and end are visible), 1·75 minutes.

The separation of the glossopharyngei did not alter respiration in any way, neither before nor after extirpation of the vagus, whether the *medulla oblongata* remained intact or was separated.

b) *Extirpation of the vagi in a marmot during hibernation, together with some observations pertaining thereto.*

As we have now learned from the highly characteristic respiration after division of the vagus to decide if the upper reflex tracts of the brain still act on respiration or not, we may find out whether, during sleep, the influence of the upper tracts continues to act or has ceased to act, and whether all or any narcotics are capable of

removing the respiratory reflexes from the brain. Single observations
which belong to this part have already been described in the previous
section; for instance, the periodic respiration which Mosso observed in
sleeping children and old people; and the periodic respiration which
other experimenters have described during the hibernation of different
animals; and the periodic respiration which some poisons cause, such
as chloral, morphia, digitalin, and muscarin.   That there are narcotics
which, during the deep sleep of animals, produce acceleration of respiration as well as a stimulation of the respiratory centre, had already
been observed by Traube with opium, when it was administered in
large doses to dogs; the same was observed by Schmiedeberg with
urethan.   But even in the sleep of hibernating animals, which can be
so deep that these animals may be operated upon and killed without
awakening, the influence of the upper tracts on respiration does not
always disappear.   I have been able to observe this in the marmot.
Unfortunately I could only obtain one specimen, as these animals
can only be caught in autumn.   When the animal arrived after its
journey it did not sleep soundly, but made single although only
weak movements of the head, and moved the paws slightly while it
was being tied and the diaphragm lever inserted.   Respiration was

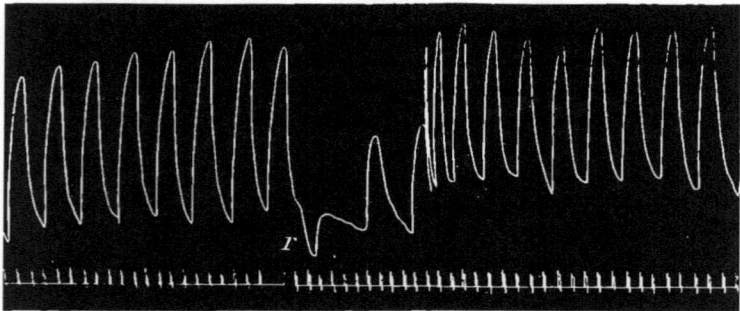

Fig. 24.—Diaphragm respiration of a young marmot during light hibernation sleep.   At *r*,
stimulation of the skin.

tolerably frequent and quite regular (fig. 24), 16 respirations in one
minute.   Skin reflexes existed and always liberated an expiratory
movement.   The lever was withdrawn, the wound sewed up, and the
animal removed to the cellar.   After a few weeks, it still slept soundly,

and did not awake when tracheotomy was performed, the vagi laid
bare and placed on threads. In performing this operation, the animal
was not tied down. This time, the respiration was registered by means

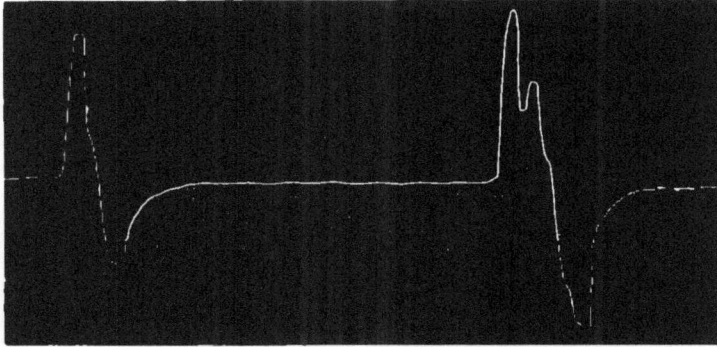

Fig. 25.—Respiration of a marmot in deep winter sleep registered by means of an air-capsule
from one of the side tubes of the trachea canula.

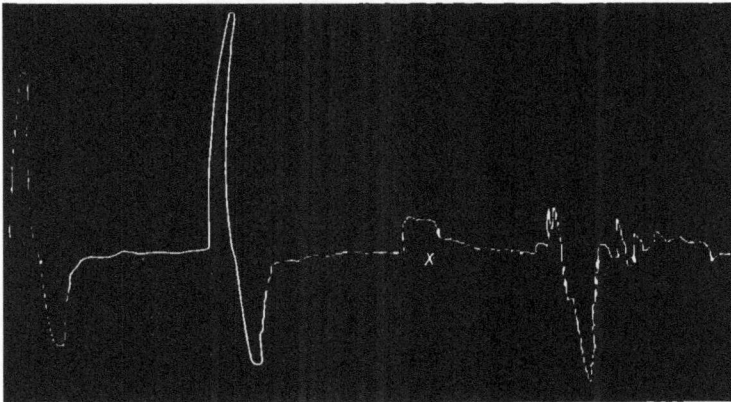

Fig. 26.—Respiration of a marmot in deep sleep during and after division of the vagi. At *x*
ligaturing the second vagus.

of an air-capsule, which was in connection with the trachea by means
of a side tube. The respiratory opening of the tracheal canula could
be narrowed when desired by means of a clip, and thus the size of the
excursions of the air-capsule could be regulated. Respiration was very

slow, not periodic but quite regular—1–1½ in one minute with active
expiration, which generally directly followed the normal passive expiration, and from which the thorax passed into a considerably long
respiratory pause in the position of rest (fig. 25). While the right
vagus was being ligatured, the animal made a deep active expiratory
movement without previous inspiration, and remained, after the tying,
in a lengthened respiratory pause (fig. 26). The second vagus was
then ligatured, and during the operation there appeared the same
phenomena as in the case of the right vagus. After extirpation of both
vagi, the long respiratory pauses were followed by more numerous
respirations—7–8 in one minute, without the animal becoming in the

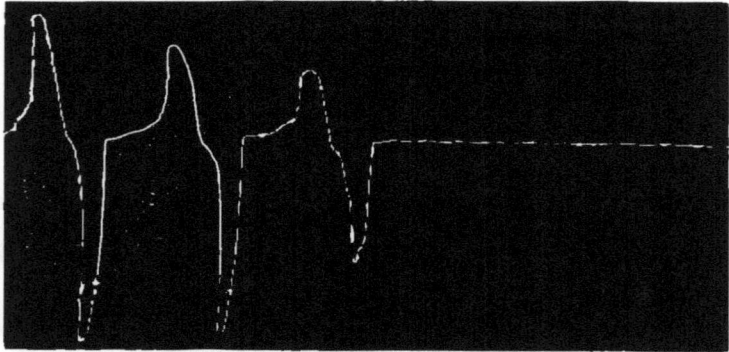

Fig. 27.—Respiration of a marmot in deep sleep after division of both vagi.

least wakeful (fig. 27). The respirations for a time were irregular:
larger respirations alternated with smaller ones, while the character of
the respirations remained the same throughout. Only occasionally, and
a considerable time after an ordinary respiratory movement had been
made without active expiration, there appeared an active expiratory
movement, while, generally, the latter directly followed an inspiration.
Gradually the respirations increased in depth, inspiration as well as
active expiration, while the frequence increased only slightly. Pinching the tail, pricking the mucous membrane of the nose, even whistling,
liberated expiratory effects and especially increased the respiratory
pause. The animal continued to sleep quietly, moved itself from time
to time a little, apparently unconsciously, and it also yawned once. Its

rectal temperature was 7·3° C. (similar to that of its kennel) while the temperature of the room was 12·4° C. Therefore, after removal of the vagi, no respiratory spasms appeared, so that it could be safely concluded, that notwithstanding the deep sleep, reflexes from the brain tract continued to act on the respiratory centre, and were able to supply the place of the vagi which had ceased to act. Then both carotids were opened and the marmot was bled to death, and finally the heart was removed. *Nevertheless respiration did not cease.* On

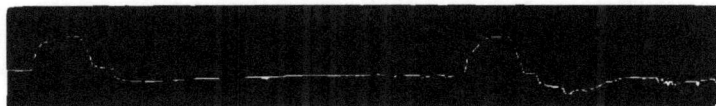

Fig. 28.—Respiration of a deep-sleeping marmot, after the vagi have been divided and both its carotids opened, and finally its heart completely separated from the body.

Fig. 29.—Beat of the heart removed from the marmot registered by the Frog's-heart Manometer.

the contrary, the respirations followed one another regularly, and this for half an hour, at the interval of a minute; only the respirations were flatter and longer, especially active expiration was less distinctly pronounced and very much prolonged (fig. 28). The heart isolated from the body continued also to pulsate for a long time—5–6 times a minute. Fig. 29 shows such pulsations, which are registered by means of Kronecker's Frog's-heart Manometer.

c) *Removal of the upper tracts after division of the vagi in the neck.*

If a healthy rabbit be taken and both the vagi in the neck divided, then, as is known, respiration becomes altered in a typical manner; it becomes slower and deeper and is at first accompanied by pronounced dyspnœa, but is throughout regular. But after tracheotomy has been performed, the well-marked dyspnœa soon disappears, respirations again become quicker, and there are no signs of respiratory spasms.

If at this time you cut transversely through the *medulla oblongata* above the respiratory centre, then suddenly the phenomena are changed as violently and quite in the same way as when at first the *medulla oblongata* was separated and then the vagi divided.

d) *Extirpation of the lower tracts.*

In order to ascertain the influence on respiration of those sensory nerves whose connections with the respiratory centre pass through the spinal cord, a rabbit was taken and the spinal cord divided at the level of the last cervical vertebra. In rabbits, division at a higher point cannot be easily performed. Rabbits bear section of the spinal cord very badly, probably owing to the quick and considerable sinking of the blood pressure. The bleeding at the operation may be very slight. In order to further extirpate the remaining sensory tracts, I have often also divided the brachial and cervical plexus on both sides, of course preserving the phrenics. When the operation was successful—and I had a large number of successful cases—then respiration remains quite normal. The phenomena are not altered in the least when we now divide the *medulla oblongata* above the respiratory centre. Respiration is frequent, regular, not dyspnœic; but division of the vagi immediately causes the characteristic respiratory spasms previously described. When the experiment is so made that first the spinal cord is divided and then the vagi, respiration can scarcely be distinguished from that caused by separation of the vagi alone, only it is more frequent—an observation which was already made by Cruikshank. When you then extirpate the brain tracts above the respiratory centre, again the respiratory spasms immediately appear in their characteristic form. Plate I. gives the whole course of the respiratory changes, in consequence of division of the different tracts, in their order. At the beginning of the upper curve, we see the diaphragm-respiration of the rabbit after division of the spinal cord at the level of the lowest vertebra of the neck. At *a* the left, and at *b* the right, vagus in the neck was divided. The lower curve shows the powerful respiratory spasms which took place after the *medulla oblongata* had been transversely divided above the respiratory centre (two curves traced through one another).

e) *Conclusions. Tonus of the vagus.*

From these experiments by extirpation we must conclude as follows:—

1. That the upper tracts are capable of replacing the vagi when these cease to act, as the vagi compensate for the cessation of action of the upper tracts; but the lower tracts are not capable of taking up the work of any of the other tracts.

2. The great difference between the individual tracts lies in this, that whilst the vagi are in constant stimulation and act continuously on the respiratory centre they also possess *tonus;* the upper tracts have only a secondary influence on respiration. From the brain flow all voluntary impulses of respiration, all emotions, all sensory impressions, and stimuli which are the result of mental actions. These all modify respiration.

3. The vagus is a sufficient, and possibly with the animal organism at absolute rest the only, active regulator of respiration, converting the inherent activity of the respiratory centre into regular respiratory movements.

C.—STIMULATION OF THE UPPER TRACTS.
HIGHER RESPIRATORY CENTRES.

It had already occurred to Budge that the higher parts of the brain had some action on respiration. By stimulation of the *optic thalami,* the *corpora quadrigemina,* and especially of the *corpora striata,* he obtained in rabbits, as already mentioned, quickening of the respiratory movements of the diaphragm, and this at the rate of from 64 to 89 in one minute, "certainly a most striking appearance," says Budge, "a deception was not possible; the changes were too strange; the respiration was tolerably regular." Sneezing shows that from the mucous membrane of the nose strong reflex expiratory movements may be liberated. Schiff found (1864) that when he stimulated mechanically and with moderation a sensory nerve of the head—for instance, the auricular nerve or infra-orbital nerve—the number of respirations decreased, and only returned to their normal number when stimulation ceased. Separation of the vagi and pulling out of the accessorii did not alter this condition. He observed further that stimulation of the superior maxillary branches of the *trigeminus*, which supply the nares of the rabbit, or the pressing of the skin of the *alæ* of the nose, slackens respiration in a marked manner, even when a canula is put into the trachea, or the superior laryngei branches of the vagi have been cut. Only one should always avoid causing pain, because this of itself increases the number of respirations. According to Schiff, one obtained the same effect by pressing the base of the rabbit's ear from behind forwards. The number of respirations then fell to one-fourth of the former number. Falk observed expiratory effects when the animals were dipped in water. Frédericq also observed this effect follow dipping the mouth of the animal in water; if he stimulated the vagus during cessation of respiration, periodical inspirations were caused. P. Bert obtained with weak stimulation of the right nasal branch of the infra-orbital nerve, quickening; with medium stimulation, slowing; with strong stimulation even cessation of respiration, in inspiration as well as expiration. Kratschmer (1870) found that stimulation of the terminal branches of the *trigeminus* causes stoppage of respiration in expiration. Division of the vagi did not at all remove this influence on respiration. Electric stimulation of the supra-maxillary branches of the *trigemini* had, although in a different and not so quick manner, a marked influence on respiration; stimulation of the *dorsalis pedis* nerve was without effect. Filehne found that extirpation of the cerebrum caused quickening of the respiration for some seconds, and attributed the same to stimulation caused by the cerebral accelerating nerves. When, however, he penetrated from the *corpora quadrigemina* deeper in the direction of the *crura cerebri* with a pointed instrument, the frequency of the respirations rose considerably (from 50 to 200!), with the occurrence of general motor movements; and the increase continued for a long time. Before this, Martin and Booker were able to establish a spot on the lobes of frogs which, when stimulated, caused inspirations and quickening of respiration, and they felt obliged to confine the position of the inspiratory centre to this spot. At a later period, Christiani studied more closely the influence of higher sensory nerves and the upper parts of the brain on respiration, and found that stimulation of the optic nerves had a quickening effect on respiration—that is to say, caused inspiration; and indeed, although the effect was not so powerful, yet it was quite like that caused by stimulation of the central end of the vagus, while the influence on respiration, through stimulation of the *trigeminus*, was quite the opposite, as it only caused expiratory effects. Therefore, the normal stimulus for the optic, light, produced on respiration, before and after removal of the brain, a quickening and inspiratory effect; and this occurrence was by far more powerful after than before removal of the brain. Mechanical stimulation of the nerves of vision and of the fifth produced the results already described. Stimulation of the auditory nerve also, according to Christiani, had an influence—always with an inspiratory effect. The excitability, indeed, for auditory reflexes showed itself to be more pronounced after removal of the hemispheres of the *cerebrum* and *corpora striata* than the excitability for optic respiratory reflexes. Further, stimulation of the upper surface of the *optic thalami* had an inspiratory effect. In the interior of the *optic thalami*, according to Christiani, there is a circumscribed spot a slight distance above the floor in the side wall of the third ventricle, close to the *corpora quadrigemina*, the mechanical, thermal, and electric stimulation of which causes stoppage of the diaphragm in inspiration. On its stimulation, respiration got deeper and more frequent. The spot is bilateral and of the size of a cubic millimetre. When this spot is stimulated, respiration becomes exceptionally quick, at the same time it becomes deepened in expiration as well as in inspiration—but sometimes only deepened, not quickened; there also appears a surprisingly lively activity of the concomitant respiratory movements (which?) and strong rhythmic accompanying movement of the tail. Double section of the vagi does not alter anything in the respiratory condition of the stimulated inspiratory centre. Mechanical destruction of the centre is very easy (what effects?). Further, according to Christiani, in the substance of the front pair of the bodies of the *corpora quadrigemina* (the *nates*), close under, and next to, the aqueduct of Sylvius, an expiratory centre exists. Its stimulation causes, before as well as after section of the vagi, the appearance of explosive active expiratory shocks, or stoppage of respiration in ordinary or in active expiration.

After the close of stimulation, a compensatory acceleration in the rate of respiration generally takes place and at the same time often a peculiar cry is heard, just the same as happens after stimulation of the *trigeminus*. The inspiratory centre in the third ventricle also remains excitable during apnœa.   After removal of the two anterior bodies (*nates*) of the *corpora quadrigemina*, Christiani, as well as Martin and Booker, was able to localize a second centre of inspiration. According to Christiani, there lie in the *pons Varolii* and *medulla oblongata* offshoots of the three upper centres, contained in the optic lobes and *corpora quadrigemina*, which can no longer be sharply separated from one another by stimulation experiments!   Steiner also thinks that he recently observed an accessory expiration centre in frogs, said to be situated about the region where Martin and Booker have located their centre of inspiration; still, Steiner does not consider his observations as conclusive, and he is of opinion that this centre is not to be identified with the true respiratory centre, which in the frog also lies in the *medulla oblongata*.   That the olfactory nerve can also exercise an influence on the movements of respiration has been settled by A. Gourewitch under the direction of Luchsinger.   On stimulating the terminations of the nerves of smell with the fumes of bisulphide of carbon, &c., and by electric currents, Gourewitch obtained not only quickening of breathing but also the individual respirations became shallower, especially after weak stimulation, while strong currents caused a complete stoppage in the position of expiration.

At the beginning of this section, I gave proof of the considerable influence which the upper tracts have over the respiration.   At the same time I demonstrated that this influence may escape notice in normal respiration, and that division of the upper tracts, when the vagi are intact, does not alter respiration in any way.   We now learn that it is especially the optic and auditory nerves that influence respiration in an inspiratory sense, the olfactory and *trigeminus* in an expiratory sense, whilst the remaining sensory nerves, according to the strength of the stimulation, and when they produce sensations of pain, are capable of producing inspiration as well as expiration.   My experiments regarding direct stimulation of the respiratory centre in the fourth ventricle have further shown that the respiratory centre can only be stimulated in the same way as if the stimuli were transmitted by centripetal nerves.   The upper tracts also act on the vagus centre like sensory nerves, and as they have no direct connection with the same they probably act through the agency of other nerve centres which are related anatomically to the vagus centre. But if one were to call every nervous distribution from which respiratory movements can be stimulated a respiratory centre, then the discovery of such new centres would have little value to science and would bring little honour to the discoverer. At the very least, one must demand from a centre, that in it centripetal tracts terminate and centrifugal tracts arise, and that its disappearance destroys or materially disturbs any function which it governs. But none of the facts which Martin and Booker, Christiani, and Steiner have placed before us afford a conclusive proof that in the upper parts of the brain there are actual centres of respiration in the same sense as has been so often demonstrated regarding the centre in the fourth ventricle. On the contrary, all the experiments of these investigators show us only that, as from the lungs and spinal cord, so also from the brain, centripetal stimulations go to the *medulla oblongata*, which act in a reflex way on the respiratory centre.

D.—CHEMICAL STIMULATION OF THE MEDULLA OBLONGATA. COMMON-SALT DYSPNŒA. DYSPNŒA CAUSED BY COLD.

A series of observations which I will attempt to describe gave rise to the opinion that it is highly probable that, after all, the phrenics have a direct connection with the brain; or, in other words, that motor influences may pass from the brain to the phrenic nerves. Therefore it is not impossible that in the cortical substance of the brain a *motor* centre for respiration may be found. When one considers how exactly voice and speech can be modulated, how singers learn to control the tones at will, and this to the finest degree, one is driven to the idea that the muscles of respiration, even when the rhythmically active centre is excluded, can be stimulated directly like other groups of voluntary muscles. Further, one must also attribute to the diaphragm a fine muscular sensation, and so it is not surprising (although at the first glance it may appear strange) that in the nerves of the diaphragm sensory fibres exist, as Budge and Panizza first demonstrated, and as Schreiber, V. Anrep and Cybulsky and Marckwald have confirmed. Ehrlich recently has been able to demonstrate these by micro-chemical

F

means. The following experiments have led to the preceding considerations:—When one stimulates the cut surface of the *medulla
oblongata* above the respiratory centre with common salt, then very
peculiar movements of the diaphragm appear, which are quite different
from those liberated reflexly from the *medulla oblongata* by means of
electric currents. For instance, the respirations which in the one case,
on account of deep section of the medulla, were intermittent and
followed one another at long pauses, immediately became very
frequent—quite irregular, and so small that one might have mistaken
them for heart-beats; but they occurred much oftener than the heart-
beats. Between these there appeared, as before, at regular intervals,
the natural high respirations; during the latter these small respirations

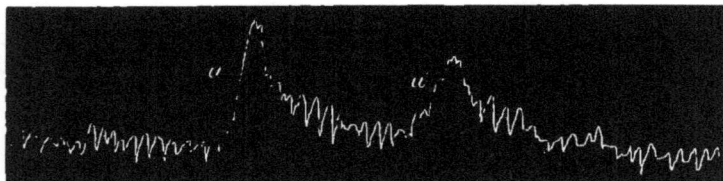

Fig. 30.   Diaphragm respiration of a rabbit during direct stimulation of the *medulla oblongata* by
means of common salt (*common-salt dyspnœa*).   At *a* and *a'*, natural respiration movements.

continued, and appeared as teeth on the ascending and descending
limb of the natural respiration curve, so that it gave quite the picture
of a blood-pressure wave curve, with the Traube-Hering waves.   I
opened the cavity of the abdomen during the experiment and could
distinctly feel the movements with my finger.   They agitated the
diaphragm in its whole extent by a continuous vibration, and as was
to be ascertained from their form, they could not in any way be
mistaken for mere quiverings of the diaphragm.   At the same time
one could feel the heart pulsating much slower.   Thus before stimulation, the number of respirations was six in one minute; during stimulation, they rose to 120 in one minute.   The height of an ordinary
respiration, on an average, was 1.6 cm., while those which were caused
by common-salt stimulation were 0.6—0.1 cm. high; and while each
natural respiration lasted one second the artificial ones lasted from
0.2—0.4 seconds.   Fig. 30 shows these respirations.

In another case, the *medulla oblongata* was cooled along the cut surface by means of a freezing-mixture of —5° C., which was allowed to run through a small double canula, having the bottom hammered out quite flat, so that at that part it was not thicker than the back of a knife of medium size. This was applied to the cut surface of the *medulla oblongata*, and could be easily inserted as far as the inner base of the skull. Before the application of the freezing-mixture, respiration was periodic. During the same, respiration at first became rhythmic, and then, during inspiration as well as during expiration, and during the respiratory pause, very frequent irregular movements of the diaphragm were introduced, so that thereby the original respiratory phases became very peculiar in appearance, and soon resembled the diaphragm quiverings which were obtained on stimulating the phrenics directly with shocks not following one another quick enough to cause a discontinuous tetanus (compare fig. 10A). At another time, the natural respirations appeared as if they were divided into three, four, or five distinct ones, while those introduced as above, and which appeared at the time of the natural inspiratory movements, sometimes reached the height of the latter. But from the rhythm of respiration, one could plainly see that we had not to deal in any way with an increased frequency of the ordinary respiratory movements,

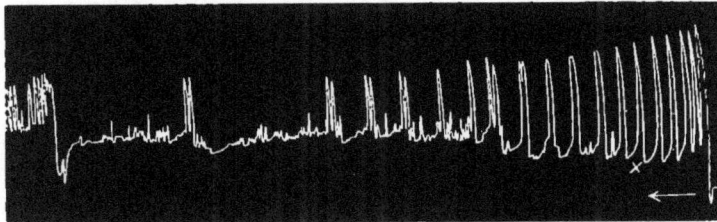

Fig. 31.--Diaphragm respirations during direct stimulation of the *medulla oblongata* by means of cold reaching —5° C. (*dyspnœa caused by cold*). At *x*, beginning of action of cold. Curves run from right to left.

but that these movements were divided by means of intercalated contractions of the diaphragm. With further cooling, the original mode of respiration disappeared completely, and at last there remained only these extremely frequent, small, irregular respirations. Fig. 31 gives a picture of such respirations after cooling of the *medulla oblongata*

(to be read from right to left). When the cooling was stopped, then
the original periods appeared as they were before the action of cold;
renewed cooling showed the same condition. Seeing that after cessation of cooling, respiration again took the original periodic character,
the vagi were divided in the neck; immediately there appeared the
well-known picture of respiratory spasms. Renewed cooling during

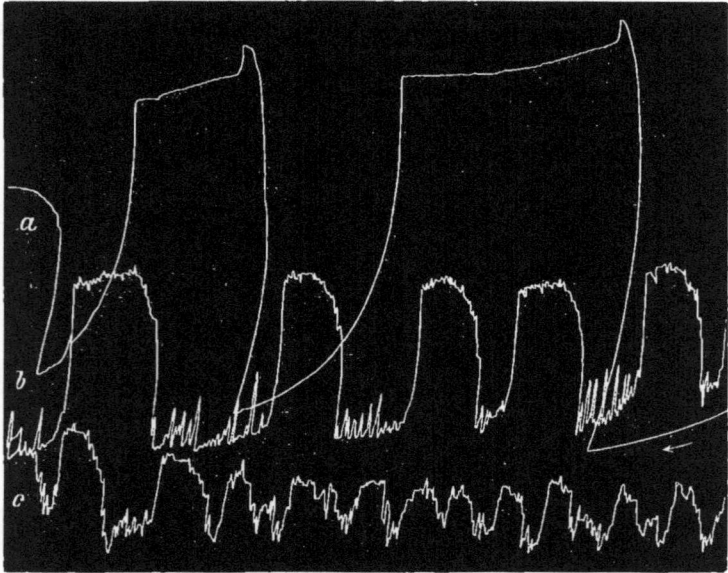

Fig. 32.— Cold-dyspnœa after division of the vagi. a, Respiratory spasm after division of the medulla
oblongata and of the vagi before the action of the cold; b and c, change of the respiratory spasms caused
by cold. Curves run from right to left.

this time had a double action: first, the respirations became frequent
and regular, and the respiratory spasms disappeared in the same way
as if the medulla oblongata had been stimulated directly with intermittent shocks; and second, there then appeared during inspiration
and expiration, as well as during the pause, the frequent small movements of the diaphragm which gradually altered the original type of
respiration, so that respiration became similar to that obtained by the
application of great cold before division of the vagus. Fig. 32, which
is also to be read from right to left, shows at line a the respiratory

spasms after separation of the *medulla oblongata* and the vagi, then at
lines *b* and *c* the results of cooling. This phenomenon, when stimulation has been caused by common salt, may be called common-salt dyspnœa (*Kochsalzdyspnœ*), and when cold was the cause, cold-dyspnœa
(*Kältedyspnœ*). The movements can certainly be produced by other
means, but however produced, they are in their form and appearance quite different from ordinary respiratory movements obtained by
reflex action from the respiratory centre. The phenomenon gives the
impression that, from the divided lateral parts of the *medulla oblongata*, motor stimulations are carried direct to the phrenic nerve with
omission of the respiratory centre. I have named these occurrences
common-salt dyspnœa (*Kochsalzdyspnœ*), and cold-dyspnœa (*Kältedyspnœ*); in the same sense as the term heat-dyspnœa (*Wärmedyspnœ*) has been used by Goldstein, Sihler, Gad, Mertschinsky, and
Richet, although a true dyspnœa is not under consideration. Thus
the diaphragm does not exhibit increased tonus, nor does thoracic
respiration occur, nor do the auxiliary muscles of respiration step into
activity. The natural type of respiration indeed remains even for a
long time quite independent of these interpolated diaphragm movements until, as it were, the latter break them up. But also in the
cephalic-heat dyspnœa (*Cephalischen Wärmedyspnœ*), which Mertschinsky has carefully described, there are scarcely to be found any
symptoms which can be described as dyspnœic. "The characteristic
of cephalic heat-dyspnœa," says Mertschinsky, "is quickening, flatten-
ing, decrease of respiratory exertion, increase in height of respiration.
It is essentially different in its appearance from dyspnœa caused by
carbonic acid, and probably also in its direct cause." It appears,
therefore, not at all improbable that central heat-dyspnœa also arises
from the same causes as common-salt dyspnœa or cold-dyspnœa.

E.—EFFECTS OF STIMULATION OF THE LOWER TRACTS. SKIN REFLEXES.
ACTION OF THE SPLANCHNIC NERVE.

During consideration of the lower tracts, we have to do mainly
with sensory nerves which act on the respiratory centre through the
spinal cord, and thus the skin reflexes become specially interesting.
We know that when one causes pain to uninjured animals by pinching

the skin, cutting, &c., besides quickening of respiration, frequent cessations of expiration occur. "Stimulation of the nerves of the limbs and of the tail," says Schiff, "increases the number of respirations. Stimulation of the anterior limbs of many rabbits leads to a lengthening of expiration and relaxation of the diaphragm; others give the same when you pinch any part of the skin." I have shown also that during sleep, in those cases where the reflexes of the upper brain tracts still continue to act, that pinching the tail and paws has as a result *expiratory* effects and stoppage of the diaphragm in expiration. But in general, in normal animals, gentle touching of the skin has no visible effect on respiration. This is different in animals which have had the brain removed. If the *medulla oblongata* is divided from the upper parts of the brain, then animals decapitated in this way are extremely excitable by reflex action from the skin. Not only pinching or the application of cold, but sometimes only blowing on the skin alters the respiration, and in different ways. In some, there appear quicker and deeper respirations, with higher or deeper (ex- and in-spiration) position of the diaphragm; in others, although very seldom—and, as it appeared to me, only when some fibres of the upper brain tracts had been left—there is a lengthened expiration; whilst in a third set—and this occurred by far the oftenest, and always when the separation of the *medulla oblongata* was complete— a lengthened inspiration.

I have already referred to the facts regarding the activity of the skin reflexes during *periodic* respiration, during the movement as well as during the pause, and that every pinch of the skin produces a complete respiration, so that with intermittent stimulations of the skin, respiration may become regular. Even when the respiration following deep section of the *medulla oblongata* has already become intermittent, one can still produce complete respiratory movements, especially at the beginning, by stimulating the skin. Still, in these cases the reflex excitability soon diminishes, and one is not often successful in producing an action immediately after the one preceding, but only after considerable pauses. If the *medulla oblongata* is transversely divided above the respiratory centre, and the vagi cut, then during respiratory spasms the skin stimulations are very effective; but now they no longer liberate complete respirations but only *quicken* the liberation of different respiratory phases, and one succeeds, during expiration, in obtaining an inspiration; but also, although not so often, during inspiration, in obtaining an expiration—the latter chiefly when the natural expiration is not far distant. Budge, who had studied the influence of the skin nerves on respiration, found that when in mammals and birds the whole skin was removed, the same phenomena took place as after division of the vagi. Soon after the operation, the frequency of the respirations diminished more or less according to the extent of skin removed, and though there was a fluctuation between increase and decrease, still the small number of the respirations was the predominating appearance. I compressed the abdominal aorta in decapitated animals (Stenson) in order to remove the influence of a part of the lower sensory nerves, and found that the rhythm of respiration was only modified in so far as it became much more frequent. After section of the vagi, respiration, as usual, became spasmodic.

In the year 1859, Budge found that stimulation of the splanchnic nerve, the sciatic nerve, or the intercostal and other subsidiary nerves caused a strong and lasting expiratory movement; and that respiration can be completely stopped. In the year 1881, Graham rediscovered this fact. The action of the abdominal fibres of the vagi on respiration has been known for a long time, as, for example, in vomiting. Recently, Knoll investigated this action minutely, and established that the expiratory movements occurred after stimulation of the ends of the vagus in the abdomen. Holmgren saw that sprinkling the abdomen and wall of the breast with water caused stronger and deeper inspirations, &c. Nevertheless, my section experiments have proved that the lower tracts, as a general rule, do not take part in ordinary respiration, that they possess no tonus, and although they cannot take the place of other tracts, and therefore cannot by means of their influence retard the respiratory spasms of the centre, still one learns from the stimulating experiments that the part played by these nerves under certain circumstances may become very important to respiration. Traube had already brought forward the influence of the nerves of the skin on respiration in explanation of the Cheyne-Stokes' phenomenon, after Volkmann had maintained that their influence on respiration was even as great as that of the vagus.

Now we finally come to the centripetal nerves which terminate in the *medulla oblongata*, and to which we must allow a specific function in the accomplishment of respiration, namely, the glossopharyngeal nerve, and the vagus, with its branches, the superior and inferior laryngeal nerves. I have examined the effects of these nerves on respiration always after the upper brain tracts had been extirpated. Only in this way, according to previous considerations, can one hope to obtain unequivocal and clear results as to the direct influence of the individual nerves upon respiration. Because one must not forget that, in the first place, these nerves are sensory, and as has also been proved by Schroeder van der Kolk regarding the vagus, they have connections with the brain, which, besides the specific stimulation, can, to a greater or less extent, influence the effect. Traube drew attention to this fact, but it has not been attended to.

a) *Effect of stimulating the glossopharyngeal nerves.*

The action of the glossopharyngeal on respiration has not been clearly shown up till the present date. All that I could find regarding it in literature is a remark of Schiff's that galvanization of the trunk or branches of the glossopharyngeal causes the diaphragm to contract, while the respirations become more constant; and a statement by Knoll, in a note of his work regarding the innervation of respiration, that stimulation of the central cut end of the glossopharyngeal causes marked inspiratory changes of the respiration. It cannot be gathered from this remark whether Knoll made experiments regarding this himself or only took the facts from literature (Schiff?). At any rate it is very difficult to bring the statements of Schiff and Knoll into harmony with a fact discovered by Kronecker and Meltzer while studying the mechanism of swallowing. In the accomplishment of the act of swallowing, according to these authors, there is first a stimulation movement and then an inhibition. During the inhibition of swallowing, the central tract of which runs in the glossopharyngeal, there takes place at the same time an inhibition of respiration, which these authorities were inclined to regard as an irradiation from the centre of swallowing to the centre of the vagus. This combination of swallowing and respiration inhibition could scarcely be understood if stimulation of the glossopharyngeal caused violent inspirations. My investigations have now shown that the statements made by Schiff and Knoll must be founded on some mistake. The glossopharyngeals in rabbits can be prepared without much difficulty. By pressing aside the submaxillary gland and dividing transversely the tendon of the stylohyoid muscle between the external and internal carotid arteries, one can see the hypoglossal nerve passing downwards and from within outwards. A little to the side of this, and between it and the trunk of the vagus, there pass, close below the styloglossus muscle and at the edge of the stylopharyngeus, two fine nerve trunks running transversely, of which the lower one is the pharyngeal branch of the vagus nerve and the upper is the glossopharyngeal. One can get the trunk of the latter so far against the pharynx and against the jugular foramen that a very fine well-isolated pair of electrodes can be passed under it without difficulty. But the nerves are very thin and tear very readily, so that it is best to use older animals, if possible, for these experiments. Now ligature the glossopharyngeals in the neighbourhood of the pharynx. As already remarked, respiration is in no way altered by this, whether the vagi had been divided before or not.

If we now stimulate the central cut end of the glossopharyngei with intermittent shocks of very low intensity (Du Bois-Reymond's sledge induction-coil, 1 Daniell, 20–30 Unit., $\frac{1}{10}''$ J) then respiration stops, and in the exact position in which it was when the stimulation began to act: during inspiration or during expiration, at the top of the first as well as after the latter had finished, and in any intermediate position. I said as soon as the stimulation began to act, because there always elapsed a considerable time ($\frac{1}{2}$ to 1 second) after the shock was given before the action took place—a time which is much longer than one sees, for instance, passing between stimulation and action of the laryngeal or vagus. In order also to see a stoppage of the diaphragm at the height of inspiration, we must stimulate during the first half of the movement of inspiration; if we wish to see the stoppage in the middle of expiration then the stimulation must take place at the apex of the inspiration; and so on. *The action of*

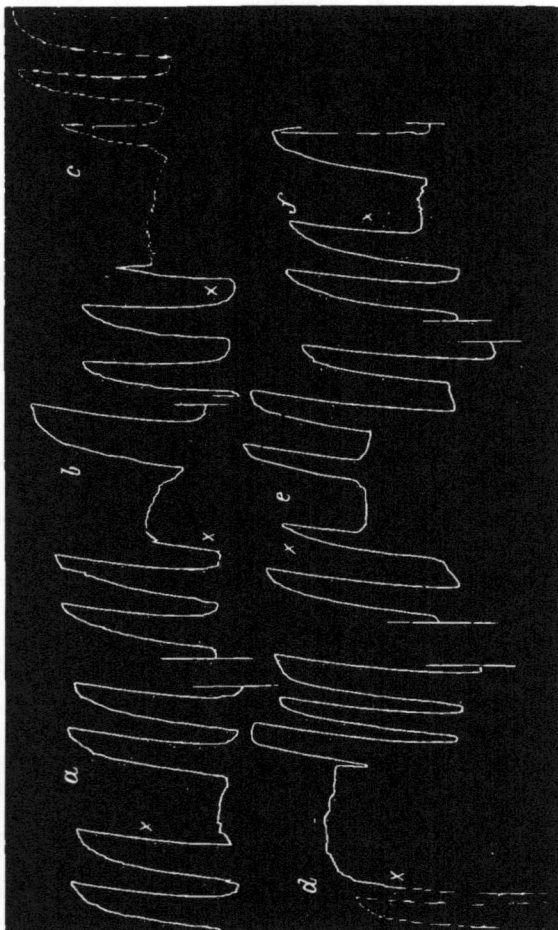

Fig. 33.—Diaphragm respiration of a rabbit during stimulation of the central cut ends of the glossopharyngeus with intermittent shocks. Six parts of curves (*a* to *f*) with stoppage of respiration in the different stages of respiration. ($\frac{1}{5}$" Interv. 20–300 Unit. 1 D). At *x* is the commencement of stimulation.

the stimulated glossopharyngeal, therefore, consists in an absolute inhibition of respiration, and lasts for about the time occupied by three preceding respirations. Then respiration begins again, even when the stimulation continues, and this with an inspiration which starts from the position in which the diaphragm was arrested at the time of stoppage. When the stimulation is continued, the diaphragm gradually, or, when the stimulation has already been interrupted, quickly, goes back to its original position of rest. In fig. 33, one sees the stoppage of respiration in the different phases of respiration after intermittent stimulation of the glossopharyngeals. Stimulation with still weaker currents interrupts the respiration, so that the diaphragm, from the moment that the stimulation becomes active, passes over quite gradually to a position of expiration. But one never obtains a slowing of respiration by stimulating the glossopharyngeals, let the currents be ever so weak, such as occurs when the superior laryngeals or *trigemini* are stimulated, but always absolute stoppage of the same as soon as the stimulation becomes at all active. The inhibition of respiration by stimulation of the glossopharyngus is, on the whole, short and cannot be lengthened, especially by strong currents; on the contrary, strong currents destroy soon the excitability of the nerves; weak currents acting frequently, or with short pauses after one another, do the same. Longer pauses restore the original excitability. One can bring the inhibitory action of the glossopharyngei especially well to the front by mechanical stimulation of the nerves. Gentle pressure or tearing, even touching with a sponge, is sufficient to bring respiration to a stand-still, just the same as by means of electricity, at whatever stage you please. The result of stimulation of the glossopharyngeals is the same whether the superior laryngeals have been divided or not. If you ligature the vagi in the neck so that respiratory spasms take place, then by means of intermittent stimulation of the glossopharyngeals, the spasm of the diaphragm can be greatly shortened, sometimes even completely interrupted, while the diaphragm gradually goes over to the position of expiration. Fig. 34 shows such respiratory spasms modified by stimulation of the glossopharyngei.

The action of the glossopharyngeals on respiration is still, after division of the vagi, an inhibitory one, and one is not in any way successful, by means of long or rhythmic intermittent stimulation, in liberating frequent and regular respiratory movements instead of respiratory spasms. In the relation of the glossopharyngeals to respi-

ration, there is a fundamental difference when compared with the
superior laryngeal, as I shall show immediately. *The glossopharyngeus is a true inhibitory nerve of respiration.* That it possesses no
tonus I have already mentioned. It comes into action during respira-

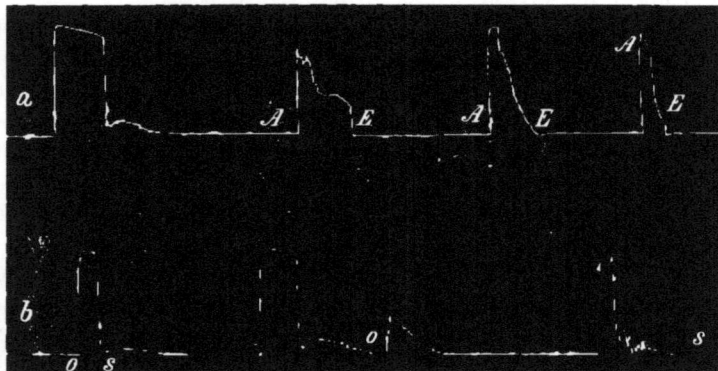

Fig. 34.--Respiratory spasms modified by stimulation of the glossopharyngeus during the different stages of the spasm ($\frac{1}{5}$" Interv. 450-500 Unit. 1 D.). At A beginning, at E end of stimulation.

tion only under exceptional circumstances, as, for instance, during the
inhibitory mechanism of swallowing; at the beginning of swallowing,
it inhibits the movement of respiration.

b) *Effects of stimulating the superior laryngeal nerves.*
Tonus of the same.

Rosenthal discovered the expiratory action of the superior laryngeal,
and he thought he had thus found the key to the contradictory results
to which stimulation of the vagus had hitherto led. He showed that
the effect of stimulating the superior laryngeal was quite the opposite
of that obtained by stimulation of the vagus itself, and he considered
the same as an inhibitory nerve of respiration. The fact that on
stimulating the laryngeals one observes stoppage of respiration in
expiration has been confirmed on all sides since the research of
Rosenthal, only V. Anrep and Cybulsky think they have proved that
in the superior laryngeal nerve there are not only expiratory but also
inspiratory fibres. But that this expiratory action was limited to the

superior laryngeal nerve was often denied, as, for instance, by Schiff, who emphasized the fact that division of the superior laryngeals neither alters the number nor the form of the respiratory movements. Further, according to Schiff, a similar effect to that produced by stimulating the superior laryngeal nerve was observed after irritation of many other nerves. P. Bert and others were of the same opinion.

In fact, when one divides both the superior laryngeal nerves in the neck (always after extirpation of the upper parts of the brain) respiration remains absolutely unaltered in height as well as in frequency. Nor with divided vagi do the respiratory spasms sustain any change after division of the laryngei. If you now stimulate the central cut ends of the laryngeus with very weak intermittent currents (1 D. 10—30 Unit. $\frac{1}{20}''$ J.) there appears, even with the weakest currents, a slowing of respiration, and a distinct lengthening of the respiratory pause. In this way, single respirations become deeper and longer, and one sees plainly the summation of the shocks; the respirations gradually increase in size, then they diminish. With stronger shocks, there is arrest of respiration, and this always in a position of expiration. This stoppage may last for a longer or shorter time according to the strength of the stimulation. I have observed relaxations of the diaphragm corresponding to the time

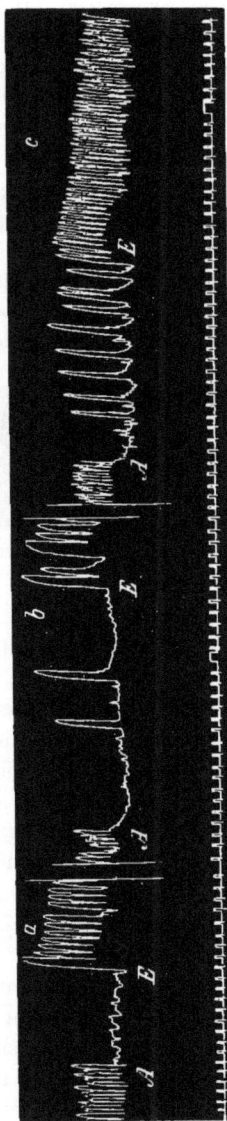

Fig. 35.—Diaphragm respiration of a rabbit during stimulation of the superior laryngeals with strong and weak intermittent currents. a, stimulation with strong shocks ($\frac{1}{20}''$ interv.); b, stimulation with medium shocks ($\frac{1}{10}''$ interv. 1 D.); c, stimulation with weak shocks ($\frac{1}{10}''$ interv.). At A, beginning, at E, end of stimulation.

occupied by sixteen previous respirations, and still longer. After stimulation, the diaphragm always increased in tonus, a fact which means that during the following respiratory movements, the relaxation of the diaphragm was not so great; also the first respirations were sometimes deeper and longer than those occurring before the stimulation; but generally they showed no difference from previous ones (fig. 35). If respiration before stimulation of the superior laryngeals was already periodic (after deep separation of the *medulla oblongata*), then by means of intermittent stimulation of the nerves, one could dispel the periods and obtain continuous respiration, during which the single respirations got interrupted, at first by short pauses, and then by pauses gradually getting longer. When the stimulation was stopped,

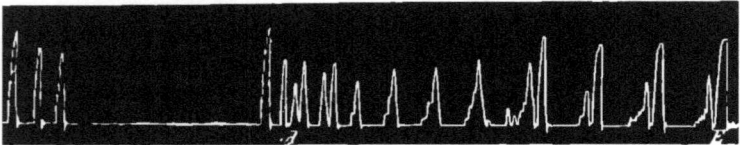

Fig. 36.—Stimulation of the superior laryngeals during periodic respiration with intermittent currents ($\frac{1}{5}$." interv. 00 Unit. 1 D.). At A, beginning, at E, end of the stimulation.

periodic respiration immediately reappeared (fig. 36). When the *medulla oblongata* was cut above the centre of respiration, and at the same time the vagi in the neck were divided and the laryngeals were stimulated during the respiratory spasms with intermittent shocks, one was successful in interrupting the spasm during inspiration and in liberating an expiration, or in lengthening this during the respiratory pause. When the stimulation was continued for a longer time then one obtained regular respiration instead of the respiratory spasms, as by direct stimulation of the *medulla oblongata*. But this differed considerably from the respiration which one obtained by direct stimulation of the *medulla oblongata*, as well as from that obtained by stimulation of the central cut ends of the vagus.

It was a rhythmic interruption of the inspiratory spasms with lengthened expiration; the pauses changed in duration between the single respirations, as is shown by fig. 37. If we divide the vagi in the neck, and at the same time stimulate the superior laryngeal nerve, then already while we are cutting the first vagus, the diaphragm

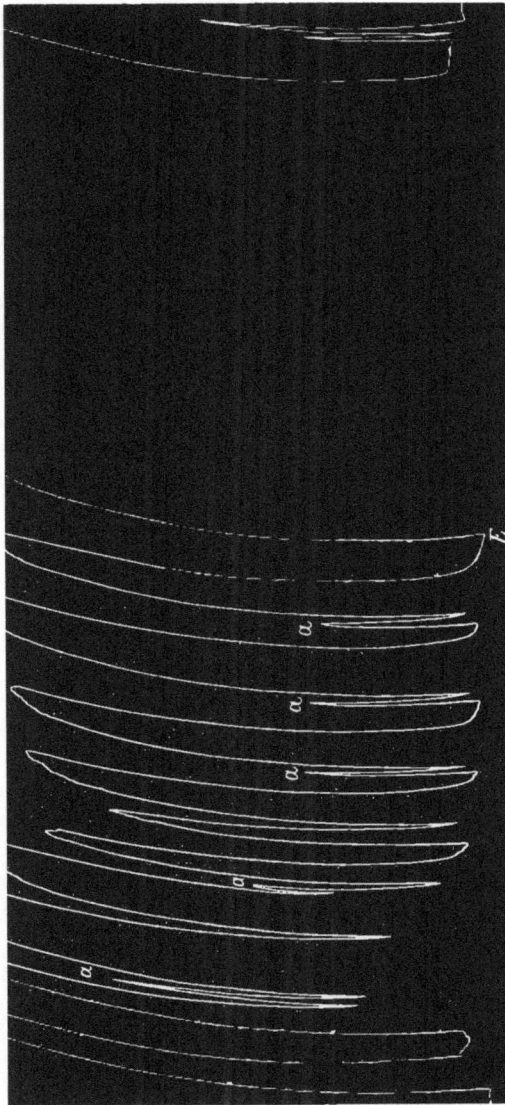

Fig. 37.—Stimulation of the Superior Laryngeals with intermittent currents during a respiratory spasm after separation of the vagi in the neck (½″ interr. 500 Unit I D.) At A, beginning, at E, end of the stimulation. α, α, show the so-called respiration of deglutition. After termination of stimulation there was a respiratory spasm.

sinks into a position of extreme expiration, and remains in that position for some time, even when the second vagus is then divided. It therefore outweighs the influence of the stimulated laryngeals so much that the effect of division of the vagus does not come into action. If the stimulation is now interrupted, the diaphragm, as usual, gradually goes over to the spasm of inspiration, and is now joined by thoracic respiration.

If we now consider the curves produced by stimulation of the superior laryngeals on fig. 36 and fig. 37 there is another fact to be observed that is well worthy of notice. One sees always during the stimulation, and as soon as an act of swallowing takes place (liberated by the superior laryngeal), that there is a short inspiratory movement of the diaphragm. This curious phenomenon constitutes the so-called respiration of swallowing—already observed by Kronecker and Meltzer, then by Steiner, and more recently minutely described by Knoll. It continues 0·4—0·5 seconds; its height varies from 2·1—2·8 cm. Besides, as it becomes more distinct after separation of the vagi (where otherwise only respiratory spasms and no short respiratory movements can be liberated), it is almost certain that in the so-called respiration of swallowing we have not to do with a proper respiration but only with a passive movement of the diaphragm, caused by the wave of contraction in the œsophagus, in consequence of deglutition, pulling down the diaphragm along with it.

The superior laryngeals are therefore true nerves of expiration, and differ in their action upon respiration from the sensory nerves of the skin in so far as, without exception and in a much more pronounced manner, they effect expiration and stoppage of respiration, and they continue to do this after the *medulla oblongata* has been separated above the centre of respiration, in which case normally the reflexes from the skin almost exclusively liberate movements of inspiration or complete respirations. The superior laryngeals further differ essentially from the glossopharyngeals in that they can only effect stoppage of respiration in a position of expiration; their action is analogous to that of the *trigemini*, and perhaps to that of the splanchnics. *In this sense one may call the last-mentioned nerves "nerves of expiration," and the glossopharyngeus a "nerve of inhibition" of respiration.* Like the *trigeminus, splanchnicus,* and *glossopharyngeus,* the *laryngeus superior* does not take part in normal respiration, and acts only as an auxiliary nerve. It possesses no tonus.

That we can also obtain expiratory effects from stimulation of the inferior laryngeal was first shown by Burkart, and since his time it has been confirmed by many experimenters in recent times, especially by Knoll. Whether the inferior laryngeals have a specific action on the respiratory centre, like that of the superior laryngeals, or whether they behave like other sensory nerves, must still be demonstrated. The inferior laryngeals are almost exclusively motor; they possess no tonus: when cut, respiration is not in any way altered.

c) *Effect of stimulating the vagi. Tonus of the same.*

Now we come finally to the consideration of the action of the vagi on respiration. Since Haller's time, the vagi have been regarded as the true nerves of respiration; but their action on respiration belongs, even at the present day, to one of the most disputed problems of physiology, and as we have seen, a consideration of their action has led to the most remarkable theories regarding the process of respiration. Traube was the first who attempted to establish precisely the mode of action of the vagi on respiration on the ground of exact experiments. He said: "There exist in the pneumogastric nerve not only motor fibres which belong to the larynx, œsophagus, &c., and sensory fibres the stimulation of which produces pain and expiratory movements (when the hemispheres exist), but also centripetal fibres the stimulation of which produces involuntary inspiration." This statement could not convince those who, when the vagi were stimulated, plainly saw *expiratory* effects as well as inspiratory, and who recognized that the assertion of Rosenthal was a mistake, namely, that the expiratory effects from the trunk of the vagus were due to unipolar stimulation of the superior laryngeal nerve. One also saw, on stimulation being applied below the superior laryngeals and when all sources of error were avoided, effects in the sense of expiration; and the view of Rosenthal, that the vagi only effected the regulation of the respiration, was not confirmed. Gad especially demonstrated that after extirpation of the vagi not only the distribution, but also especially the force and capacity, of the respirations altered. The warning of

Traube, that one could only test the action of the vagi on respiration after the upper tracts of the brain had been divided, unfortunately was neglected. They searched for all kinds of complicated explanations in order to account for the fact that in the one nerve there were two kinds of fibres acting in quite opposite modes, and Miescher-Rüsch still demands from a complete and satisfactory theory of normal respiration "that in it there should be an explanation not only of the inhibitory and expiratory, but also of the *inspiratory* reflex of the vagus." The Hering-Breuer automatic-lung-theory, as well as the inhibitory theory of Gad, are not capable of removing this difficulty in explaining the phenomena. Further, the observation made by Meltzer, which could not be confirmed by Anrep and Cybulski, that the stimulated vagus of the male rabbit has an inspiratory effect, and the vagus of the female rabbit an expiratory effect, does not give a satisfactory explanation. Knoll has recently tested the terminations of the vagus with regard to their relation to respiration, and divides the same, in the rabbit, into two kinds of fibres: (1) the one set inhibits respiration in the position of expiration; they branch partly from the cervical, partly from the thoracic portions of the vagus, and go to the larynx, trachea, and pulmonary plexus. (2) The others cause inspirations; they go from the *rami trachealis inferiores et pulmonales* to the organs in the thorax. In the dog and cat, the abdominal vagus also receives inhibitory fibres of expiration.

My experiments on the stimulation of the central cut ends of the vagus on over sixty rabbits have all been carried out after the *medulla oblongata* had been transversely divided. If you then stimulate both the central cut ends of the vagus, at the same time, with moderate and strong induction shocks, the following phenomena may be observed:—If the cut has so passed through the *medulla oblongata* that the *alæ cinereæ* have remained quite uninjured, and if the shocks take effect only during the natural respiration spasms, then in many cases one can liberate an inspiration during the position of expiration and an expiration during the position of inspiration.

So even isolated closing induction shocks, judiciously used, may in this way cause respirations at any wished-for rate. The liberation of an expiration is often immediately followed, without renewing the stimulation, by a new inspiration. In other cases, with medium cur-

rents, one is successful almost without exception in liberating with a
shock an inspiration during the position of expiration; but during the
position of inspiration an expiration can only be liberated when *several*
single shocks are given at short intervals, or when the time of occurrence of a natural expiration is not far distant. If one then allows
artificial respiration to act for a time without reaching apnœa, he will
be successful (soon after suspension of respiration) with single opening
and closing induction shocks in liberating a complete respiration,
or sometimes two respirations immediately following one another.
During apnœa single shocks have no effect. Further, if during sepa-

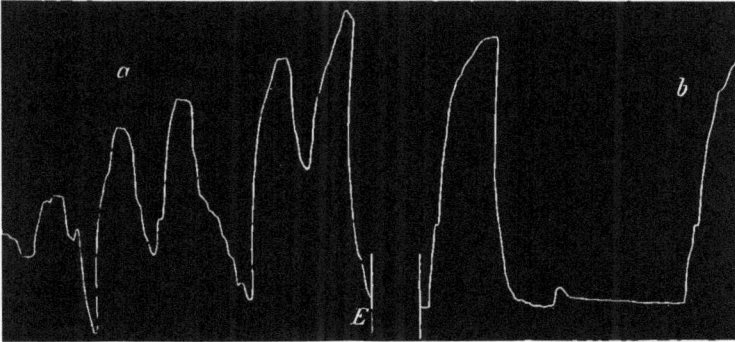

Fig. 38.—Artificial electrical diaphragm respiration in a rabbit (which had been bled) by means of
stimulation of the central ends of the vagus with intermittent currents ($\frac{1}{5}$" interv.). *a*, long-continued
action of the currents; effect of summation; *b*, rhythmic tetanizing stimulation (every 3 seconds),
alternate, rudimentary, and complete single respirations. A, beginning, E, end of the stimulation.

ration of the *medulla oblongata*, a partial injury to the respiratory
centre has taken place—the respirations following at long pauses being
short, deep, and spasmodic,—sometimes one is still successful, by means
of strong induction shocks, in liberating a respiration during the
respiratory pause. Sometimes this abdominal respiration is preceded
by a thoracic respiration. One can accordingly obtain from the vagus,
by means of reflex action, with single shocks in the case of animals
which have remained very excitable, single, deep, and long respiratory
movements; in other cases, sometimes an inspiration, or sometimes an
expiration is obtained, overbalancing inspiration. But the expiration
liberated by stimulation is never active. When the central cut ends
of the vagus were stimulated with opening induction shocks following

one another quickly (for instance in $\frac{1}{20}$—$\frac{1}{16}$ sec. interval), then one was always successful in liberating each time a deep and long respiratory movement by reflex action even in those animals which no longer breathed automatically, but in which the separated *medulla oblongata* remained without injury to the floor of the fourth ventricle; as, for instance, after bleeding from the basilar artery. This occurred even after tetanus with medium currents (compare Fig. 38, *a*). When one allowed the stimulus to act for a considerable time, then a series of respiratory movements were liberated which gradually increased in size (summation of the shocks).

Such respirations are shown on fig. 38, A. In these cases also one

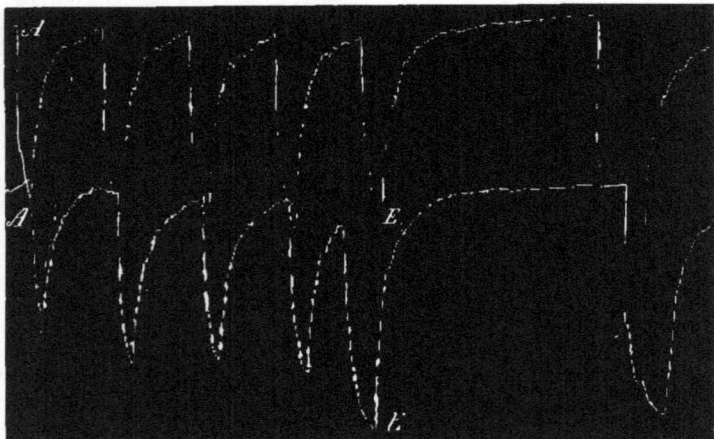

Fig. 39.—Reflex liberation of respiration by means of rhythmic (each 2″ long) tetanizing stimulation of the central vagus ends ($\frac{1}{15}$″ interv.). A, beginning, E, end of stimulation. Without stimulation, respiratory spasms 6 per minute; with stimulation, 30 respirations per minute.

sees the preponderance of the inspiration, in so far as at the beginning, the expiration is interrupted half-way from the next inspiration, till at last a complete expiration takes place, but without active expiration. If one allows the tetanizing shocks to act without pause during natural respiration, then with short duration of the stimulation, one is always successful in liberating an inspiration; and during the position of inspiration in liberating an expiration, especially when the time of the natural expiration is not far distant. Weaker tetanizing shocks, kept up for a considerable time, can strengthen the natural inspiration so much that the animal in vain attempts to expire, and gets into such a state of dyspnœa that the thoracic respiration steps in to assist. During apnœa, by means of tetanizing shocks of short duration, one can always obtain, although very incompletely, a respiration rather unlike a normal one. The most favourable conditions with regard to reflex stimulation of respiration naturally appear on rhythmic tetanizing stimulation of the central ends of the vagus.

When the animal breathes no longer automatically, as after bleeding, then each period of stimulation represents a respiration, which at first is small, gradually increases till the liberation of a respiration normal in depth and length takes place, after which the series recommences (fig. 38 b). During natural respiration one may be successful in producing deep and long respirations following one

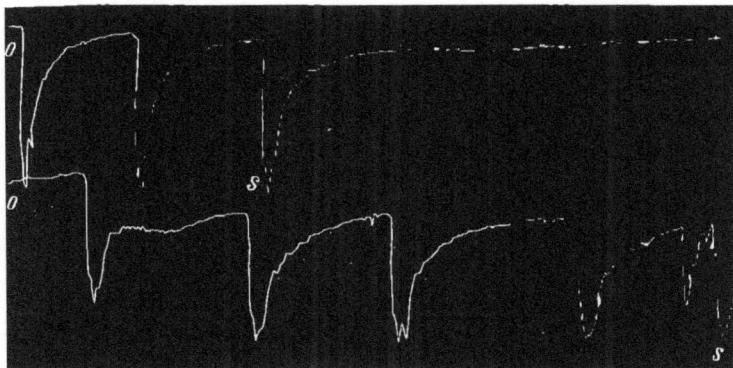

Fig. 40.—Reflex liberation of respiration by means of rhythmic (each 3″ long) tetanizing stimulation of the central cut ends of the vagus ($\frac{1}{30}$″ interv.). o, beginning, s, end of the stimulation. During the stimulation 6 respirations per minute, before and after stimulation about 1 respiration per minute.

another regularly, as they are seen on fig. 39 and fig. 40, if the requisite gradation of the shocks as regards intensity and duration be obtained; for instance, by the application of a Du Bois-Reymond sledge induction-coil (about 4000 turns of the secondary spiral with a slide for the coils of 20-15 cm.) used in connection with two Daniell's elements, while the equalizing arrangement of Helmholtz is employed to make the opening and closing induction similar in intensity, and the Kronecker interruptor is in connection with the primary coil regulated to 20 or 26 vibrations per second, so that in each period of 3 seconds there is a very short stimulation. When, in this way, by means of electrical stimulation, the animal has been allowed to breathe by reflexly acting stimulations for a considerable time, then the respiration spasms immediately following the stimulation (which are the natural respirations made by the animal itself after the *medulla oblongata* and vagus have been divided), are often so modified as to appear like those obtained by stimulation, and it is only after some time that the natural respiration spasms again occur (fig. 41).

But generally one sees inspiration and expiration spasms appearing in the old way immediately after rhythmic stimulation has ceased, until the regular artificial electrically-liberated respiration is introduced. Further, by deep section of the *medulla oblongata* (and therefore incomplete respiration interrupted by long pauses) one can liberate, during these pauses, by means of rhythmic electric stimulation, artificial respirations between the natural ones, as well as strengthen the natural respirations. When respiration after division of the *medulla oblongata* was periodic, then, as above mentioned, the periods disappeared after division of the vagi, and they could not again be obtained by rhythmic electric stimulation of the central cut ends of the vagus. In these circumstances, regular respiration sets in in the usual manner.

During apnœa, one obtains by the same stimulation, quite superficial, long-protracted movements of the diaphragm, during which it gradually goes over to the position of inspiration (increased tonus). These movements then increase in size so that they pass over to deep and long regular respirations at the time of the appearance of natural respiration. In the course of rhythmic or continuous stimulation of the central cut ends of the vagus, there never appeared active movements of expiration, but they disappeared if they existed before. As is seen on figs. 40 and 41, artificial respiration caused by rhythmic tetanizing shocks runs so that the single respirations follow one another without any pause, and the expiration is very short, as in the natural respiration of the animal. This is the rule. An exception is found in those cases in which the natural respiratory spasms were already interrupted by long pauses (a considerable time after separation of the

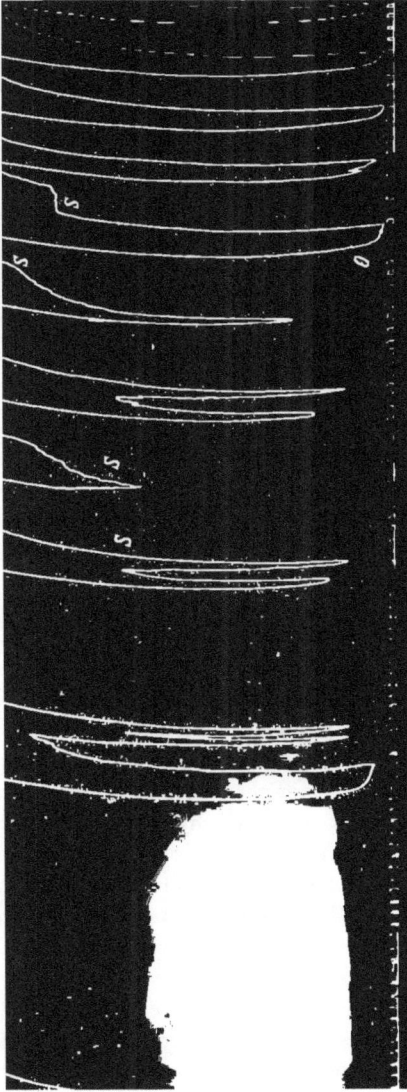

Fig. 41.—Modification of the respiratory spasms after long action of rhythmic tetanizing stimulation of the central ends of the vagus. O, beginning, S, end of the stimulation (γ'' interv. 50 E, 2 D.).

medulla oblongata, deep separation, &c.). In these cases there also appears with artificial electric respiration a pause between the single respirations, therefore expiration appears to be lengthened; but here also it is never active, as is shown by fig. 42. But otherwise we never

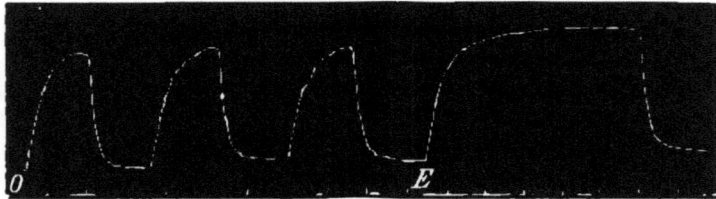

Fig. 42.—Respiration with expiration type in consequence of intermittent (every 3″) totanizing of the vagi ($\frac{1}{180}$″ interv.). o, beginning, E, end of the stimulations.

succeeded in lengthening the expiration or the respiratory pause by means of electric reflex stimulation of the central cut ends of the vagus, while the liberation of expiration, and especially of inspiration, was quickened. This condition of the vagi, therefore, was in every respect different from that of the laryngeals. With electric stimulation of the laryngeals, not only was inspiration always and immediately interrupted and an expiration liberated, but expiration and the respiratory pause could also be considerably lengthened; and while the respiration liberated by rhythmic electric stimulation of the laryngeals consisted of short inspirations, long expirations, and of respiratory pauses, varying in duration, it is quite the opposite with the respiration obtained by stimulation of the central vagi, which consists of long inspirations and short expirations, and the respiratory pause generally altogether disappears. Respiration in consequence of rhythmic electric stimulation of the vagi corresponds, when the vagus roots are not destroyed, exactly to that which appears on direct stimulation of the *medulla oblongata;* only with stimulation of the vagi, one is more successful in regulating the shocks and in maintaining respiration, which is more frequent and more like normal respiration.

What are we bound to conclude from these experiments? (*Firstly*) That the trunk of the vagus contains no expiratory fibres, and that Traube is partly correct when he attributes the action in this sense to the sensory fibres of the vagus, which, in connection with the sensory tracts of the brain, produce, through these, a sensation of pain and therefore expiratory movements. (*Secondly*) According to my experiments, the vagus contains no proper fibres of inspiration. By means of stimulation of its central end, one can liberate inspiration as well as expiration, although inspiration predominates. The respiratory centre, liberated from its centripetal tracts, so as to be only in connection with the phrenics, is, as I have shown, automatically active; but it is not capable of liberating *normal respirations*, but only *respiratory spasms*. *It requires stimulation from the periphery, so that the stimulations gathering in it are brought to a rhythmic discharge. This in normal respiration is accomplished by the vagus.* The vagus is in constant stimulation. It possesses tonus, as the section-experiments have taught us. The vagus, according to this, is constantly carrying stimulations to the respiratory centre, by means of which the liberation of inspiration and expiration is facilitated. That continuous stimulations cause rhythmic discharges is a general property of normal nerve centres. I require only to mention the rhythmic movement of the leg of the reflex frog as a result of continuous stimulation of its skin with acid (see fig. 17, p. 38), and to allude to the other examples given at that place. The normal spinal cord answers every continuous stimulation with *rhythmic* movement, while when the spinal cord is poisoned by strychnine the *accumulation* of internal and external stimulations is favoured, and wide-spread spasms are thus excited. In the same way, the respiratory centre, *liberated* from its centripetal nerve tracts, is capable of liberating only respiratory spasms. *In connection with the vagi*, it converts continuous stimulations into rhythmic movements. The vagus is therefore neither a nerve of inspiration nor a nerve of expiration. It is not a regulating nerve, as Rosenthal thought, and far less an inhibitory nerve of respiration, as was Gad's opinion; but it is a *discharger which prevents the tension in the respiratory centre increasing to an unnatural extent.*

d) *Innervation of the respiratory muscles of the thorax.*

During this exposition, I have several times remarked that under different circumstances—when the decapitated animal with divided vagi has become strongly dyspnœic, or during prolonged inspiratory-tetanus, or with pressure of a blood clot on the fourth ventricle, or with increasing dyspnœa shortly before death—thorax respirations take place, whilst the position of diaphragm is that of inspiration. We shall now inquire whether and by what means the thorax muscles of respiration can be set into action by the vagus. Accordingly, in animals with the *medulla oblongata* separated above the respiratory centre the diaphragm was paralysed by division of the phrenics. Respiration then became exclusively costal, but the type remained the same—long inspiration spasms and short expiration spasms, followed, however, by long respiratory pauses.   When the central ends of the vagus were stimulated with tetanizing shocks, one was then successful, during the position of *expiration of the thorax*, in liberating an inspiration, and sometimes an *expiration*, during the position of inspiration.   This occurred with the coils at the same distance apart as before division of the phrenics.   But, as in abdominal respiration excited by means of intermittent stimulation, to obtain rhythmic respiration, not only had the currents to be strengthened, but each period of stimulation did not liberate a movement of respiration.   To accomplish this, *two* periods of stimulation were often necessary.   *It was also ascertained that the subsidiary muscles of respiration were more difficult to stimulate by reflex action than the diaphragm.*

V

CAUSES OF STIMULATION OF CENTRE OF
RESPIRATION.

At the beginning of Chapter IV. an historical review was given of the most important opinions regarding the causes of respiration. Unfortunately we came to the conclusion that none of the proposed views could be regarded as a complete and satisfactory explanation. By our investigations, the great problem of this all-important function of life will not entirely be solved. Still it may not be useless to narrow the range of possible opinions. If we overlook Haller's theory, according to which the active mind stimulates respiration, then we must next consider those hypotheses which attribute to the blood the power of exciting respiration. Haller, who was of opinion that the individual parts of the heart are stimulated to contraction in the order in which they are filled by the circulating blood, has not allowed the blood to be regarded as the "*primum movens*" of respiration, because the fœtus does not breathe, although its circulating blood plays round the nerves of respiration for a long time. Haller's theory regarding the stimulation of the heart was given up because it was found that the excised heart, free from blood, was capable of continuing to beat for a long time. Does not then anyone ask: Is respiration not possible without blood?

Many facts show that the central nervous system—therefore also the centre of respiration—soon loses its powers of life when the blood circulation is interrupted. In the course of this work I have myself made a series of statements regarding the results of anæmia of the *medulla oblongata*. But those who assume that for each short series of respirations a new impulse on the part of the blood is necessary, would require to prove that animals in which the blood is removed always and immediately stop breathing. Now this is not the case. Not only frogs and other poikilothermic animals can breathe for a long time after the heart has been removed from them, but also, under some circumstances, one may see respiratory movements in mammals going on for an hour or an hour and a half after the circulation had completely stopped. *"Etiam evulso corde potest respiratio superesse"* was already remarked by Haller on the strength of experiments by Pechlin. Traube therefore distinguished in his "Lectures" direct "heart-death" and "respiration-death." The experiments of Volkmann, fully described at page 54, give a direct experimental proof that the circulation is not indispensable for the prolonged continuance of respirations. The kitten breathed in a regular manner for 40 minutes even after both lungs were ligatured, and thereby the circulation completely interrupted. I myself have occasionally been able to make similar observations. Thus, in a dog which, after a blood-pressure experiment, had been killed by a prick in the co-ordination centre of the ventricle of the heart, I saw respirations continue, although the heart showed fibrillar contractions and artificial respiration had been interrupted. The arterial system was then injected with paraffin from the carotid, and the injection, as the autopsy showed, filled all the main branches. Notwithstanding this, the dog showed respirations for several minutes, at long but regular intervals. I have often had occasion to see the same in animals which had been bled (in dogs as well as rabbits), especially when they had been tied down for a considerable time. In such cases, I also noticed, as mentioned above, that when spontaneous respiration in bled animals had disappeared, through stimulation of the central ends of the vagus in the neck, respirations of a regular kind could be reproduced (compare p. 101). More recently, I had occasion to make an experiment, which showed very evidently that the respiratory centre could act regularly without circulating blood. It was that of the marmot already mentioned (page 72), which at the end of the experiment was bled from both the carotids till no blood flowed from them. Nevertheless, the respirations, as already mentioned, remained almost unaltered for some minutes; gradually they became feebler and less frequent, but remained at the rate of about one respiration in the minute (almost the same rate as they were at the beginning of the experiment) for about half an hour. Even although the heart was removed, respiration did not alter to any considerable extent.

Still less necessary than the blood for respiration is the oxygen

contained in the same. Kronecker and MacGuire found that the heart
of the frog pulsates just as powerfully with blood deprived of its
gases as with that containing oxygen; while the blood of asphyxia,
or blood containing reduced hæmoglobin, soon stops its action. This
observation induced Kronecker and Sander to commence a series of
albumin-transfusion experiments with dogs and cats, in order to ascertain their influence on respiration. These experiments led to the
well-known life-saving transfusions of common-salt solutions. The
results which concern the accomplishment of respiration have up to the
present not been made known. With the permission of Professor
Kronecker, I will now communicate the most important of the experiments, according to his reports. Dogs sustained the substitution of $\frac{3}{4}$ to
even $\frac{4}{5}$ of their blood by 0·6% solution of common salt. Respiration,
which, after bleeding, had ceased before the transfusion, began during
the latter, and increased to such an extent that often after a few
minutes, and always on the next day, it was normal. Solutions of
syntonin had the same effect as solutions of common salt; if anything,
the animals seemed not to bear the syntonin so well. Solutions of
peptones acted, as was to be expected from Fano's reports, very in-
juriously, even deadly. On the other hand, by the injection of serum
from the dog or horse one could replace a still larger quantity of
blood, without injury to the animal, than with the common-salt solution; even then respiration remained normal. But especially striking
is the experiment of Von Ott, who withdrew $\frac{1}{2}$ths of the blood
from a dog and replaced the same with serum from the horse, free
from blood corpuscles. The numbers of blood corpuscles in samples
of blood taken from the larger veins of the living dog, on an average,
on the field of a quadrate micrometer, were as follows:—

Before removal of blood	.	.	.		65·40	blood corpuscles.
1 day after bleeding and serum injection					1·17	,,
3 days	,,	,,	,,	,,	1·60	,,
5 ,,	,,	,,	,,	,,	2·82	,,
7 ,,	,,	,,	,,	,,	4·13	,,
10 ,,	,,	,,	,,	,,	13·17	,,
13 ,,	,,	,,	,,	,,	17·11	,,
16 ,,	,,	,,	,,	,,	24·55	,,

This table shows that by the horse's serum a large number of the blood corpuscles of the dog were still destroyed. This result agrees with the observations of Landois. The tolerably large quantity of bile pigments in the urine during the first days after the transfusion led also to the conclusion that decomposition of hæmoglobin had taken place. During the first days after the serum-transfusion, the dog possessed $\frac{1}{35}$th of its red blood corpuscles. But the oxygen of the blood is contained almost exclusively in the blood corpuscles, as has been proved by Pflüger, who demonstrated that the serum of the dog contains only 0·1 to 0·2 volumes per cent of oxygen, whilst arterial blood contains 15 to 20 volumes per cent. Therefore, Von Ott's dog possessed only a 55th part of its oxygen, while, according to the determinations of Setschenow and Holmgren, the blood of a suffocated dog still contains 3% of oxygen, or about $\frac{1}{6}$ of the blood is still saturated with oxygen. *But this dog, with the serum of the horse in its vessels, from the first day did not suffer from distress of breathing;* it was at first somnolent and weak, but on the third day it was in its normal condition. With these results, it cannot be said that great deficiency of O causes distress of respiration, or that a normal fluctuation in the amount of oxygen causes the normal impulses of respiration. On the other hand, the impulse to respiration caused by the presence of carbonic acid, according to this theory, cannot explain normal respiration, because without a circulation of the blood, the exchange of gas in the respiratory centre can scarcely be presumed to exist. Dyspnœic spasms, therefore, must be produced by anæmia. Against this view, however, Gad, in his most recent contribution, correctly adduces the fact that by bleeding respirations are made considerably slower, and, indeed, do not pass into a state of inspiratory tonus, but the thorax returns to a position of rest.

The alternate expansion of the tissue of the lungs, which, without doubt, is capable of irritating the ends of the vagus by alternate stimulations, as the famous Hering-Breuer experiments have shown, is just as surely unnecessary for the accomplishment of the regular rhythmic respiration, because, as already Haller had adduced (and every experimenter can confirm this), animals in which the pleural sac has been opened still make regular respiratory movements. Further, the often-recorded experiments of Volkmann teach that they continue even after separation of the lung.   What then, among so many opinions that must be negatived, can we find to justify a positive statement?

In Chapter III. it has been shown that impulses of the respiratory centre still remain even after the centripetal innervation tracts have been separated.   But then the type of respiration is quite abnormal; it is spasmodic.   Therefore there must be regular impulses (which, as already mentioned, do not require to be rhythmic) kept up by excitations travelling along sensory nerve tracts.   It has also been shown that in this way, and in the first place, the terminations of the vagi in the lungs come into action, and, in the second place, higher parts of the brain.   The self-acting stimulations on the respiratory centre may be of the same nature as those which stimulate the isolated heart; perhaps the products of the decomposition of the intercellular fluids.   This view is supported not only by the most recent experiments of Langendorff, in which he proved that the brain substance, by action, becomes acid, but also by the fact that strong stimulations of the respiratory centre can be modified, but very slowly, by a complete ventilation of the latter by the circulating blood, when restitution processes in the tissue may well have had time to take place.

VI

APNŒA.

Artificial respiration seems first to have been accomplished by Vesalius in so far as he forced air into a dying animal, and in this way restored the heart-beat. According to Burdon-Sanderson, Hooke was the first to demonstrate apnœa in dogs, in October 1667, before the Royal Society of London. Hooke opened the breast of a dog and dilated the lungs by means of a bellows. In order to maintain a constant current of air in the lung, he punctured the latter on its surface. He found that "although the eyes of the animal were very lively the whole time, and the heart beat regularly, the respiratory movements stood quite still." One of the most important facts supporting Rosenthal's theory of respiration, that the deficiency of oxygen in the blood was the stimulating cause of the respiratory movements, lay in the following experiment: by means of methodic insufflation of atmospheric air Rosenthal produced a state of breathlessness, which continued for a considerable time (five minutes or more), and which could not be interrupted by stimulation of the central vagi. During this condition, which Rosenthal designated *apnœa*, the blood was said to be supersaturated with oxygen and accordingly the natural stimulations of the respiratory centre ceased. This view of Rosenthal's found many opponents. The first explanation offered by Pflüger was, that by increased tension of the oxygen during apnœa an opportunity was given for a more complete decomposition of the reducing material in the tissues, so that afterwards, from the deficiency of such material for the time being, less oxygen was consumed. This theory could not be upheld by Pflüger himself nor by his pupils. Thiry had already found that the inspiration of a mixture of equal quantities of air and hydrogen caused stoppage of respiration. Hering, who examined the gases of the blood during apnœa, found in the arterial blood of apnœic cats on the average not more, rather less oxygen than in normal-breathing animals, but the carbonic acid was considerably (to the half) lessened. Ewald proved that although apnœa implies saturation with oxygen of the arterial blood, saturation of the tissue with this gas cannot be obtained in this way. The consumption of oxygen during apnœa is neither greater nor smaller than in the normal condition. The venous blood, on the other hand, contains much less oxygen than the normal blood, so that when one considers that the greater part of the whole blood is venous blood, and that the arterial blood contains very little more oxygen than in the normal condition, according to Ewald, one comes to the "paradoxical conclusion that during apnœa the body is poorer in oxygen than during the normal condition." Filehne observed also that not only the venous blood, but also the arterial blood becomes darker than the normal before the first respiration appears after apnœa. Other considerations were also urged against the theory of Rosenthal. Brown-Séquard was of opinion that after division of both vagi apnœa could no longer be produced by insufflating air, and he considered the apnœa as a reflex inhibition caused by mechanical stimulation of the vagus. Rosenbach and Filehne also could, in animals on which vagotomy had been performed, produce rest from respiration only with much more difficulty and for a shorter time than in animals in which the nerves had been left untouched. Further, Knoll observed that in animals which were uninjured the respiration after apnœa only reappeared when other organs already showed distinct appearances of dyspnœa; for instance, when there was slowing of the pulse, increase of blood pressure, and movements of the intestine. The apnœa, according to Knoll, diminishes the excitability of the respiratory nerve centre. When he performed the Kussmaul-Tenner experiment with animals in the condition of apnœa, then, in opposition to Rosenthal, he observed no respiratory movements, but respiratory spasms, which appeared spontaneously. This observation also was in favour of the view that a diminished excitability of the respiratory centre existed. If the vagi were divided during apnœa, then, as Gad observed, no respiratory movements took place; but there only appeared passing reflex irritation. On the other hand, Knoll also saw that after division of the vagi apnœa was only seldom complete. According to him, to have complete apnœa the integrity of at least one vagus is necessary. But one can, as Knoll thinks, replace division of the vagus by means of artificial stimulation of the

H

vagus; and this is said to cause a diminution of the excitability of the respiratory centre. Miescher-Rüsch found that, as far as apnœa had anything to do with the gases of the blood, the carbonic acid alone of the blood can be considered as an active agent. He distinguishes (1) an *apnœa vera*, which, after extirpation of the influence of the vagus, appears as complete stoppage, or even only weakening, of the respirations, and arises from decrease of the carbonic acid stimulation in the respiratory centre; and (2) *apnœa vagi*, an inhibitory reflex originating in the fibres of the lung. The latter, according to Miescher, is only a special case of expiratory arrest of respiration, such as may be obtained from stimulation of the *laryngei*, of the *trigeminus*, &c., and which may be collectively termed *apnœæ spuriæ*.

After obtaining the well-known respiratory spasms in a rabbit by separation of the *medulla oblongata* above the respiratory centre, and by ligaturing the vagi in the neck, if artificial respiration is carried on in a very complete manner, these insufflations may be continued for half an hour and longer without obtaining apnœa, although by this time the blood without doubt is saturated with oxygen. During the whole time, however, of artificial respiration the respiratory spasms continue, and are manifested in a very peculiar and characteristic manner. At the time of the natural expiration and of the respiratory pause the diaphragm follows the insufflations without offering any opposition, and responds to its rhythmic movements. But as soon as the inspiration spasm takes place, the diaphragm remains contracted just as without artificial respiration; and even very complete insufflations are not capable of overcoming the contraction, but only produce flat excursions of the diaphragm in the position of spasm. As soon as the spasm is over, the diaphragm again follows artificial respiration without resistance. The form of this peculiar respiration is seen at fig. 43. This play between spasm and passive movements of the diaphragm continues during the whole time of artificial insufflation without change, although the ventilation of the *medulla oblongata* must certainly be regarded as complete. If one interrupts the artificial respiration, then the original respiratory spasms immediately reappear unchanged. For a long time I thought it was absolutely impossible to make such animals apnœic. That is not correct. One can manage, with specially complete and, at the same time, very frequent, long-continued artificial respirations, to produce apnœa even in decapitated animals, and animals upon which vagotomy has been performed. We then see the inspiration spasms grad-ually becoming shorter, until at last they completely disappear, and the diaphragm finally follows the in-sufflations without resistance. Fig. 43 shows such spasms gradually becom-ing shorter. If one continues to ven-tilate for a short time, and then in-terrupts the artificial respiration, the animal is apnœic, and remains a minute or so in the position of respiratory rest; then respiration begins again immediately, first with smaller respiratory spasms, which quickly increase in size, but never with regular respiratory movements. During apnœa the reflexes from the skin remain without effect on the respiratory centre. If one stimu-lates during apnœa the central cut ends of the vagus in the neck, then, as already described, page 99, single shocks have no action, while intermittent shocks often liberate very flat and long-protracted move-ments of the diaphragm, which are not similar to the normal respira-tory movements. These respiratory spasms during artificial respiration show that the impulse in the re-spiratory centre is not regulated quantitatively according to the change of gas. But since, by very long-continued and complete venti-lation, the spasms can be caused to disappear, one must assume

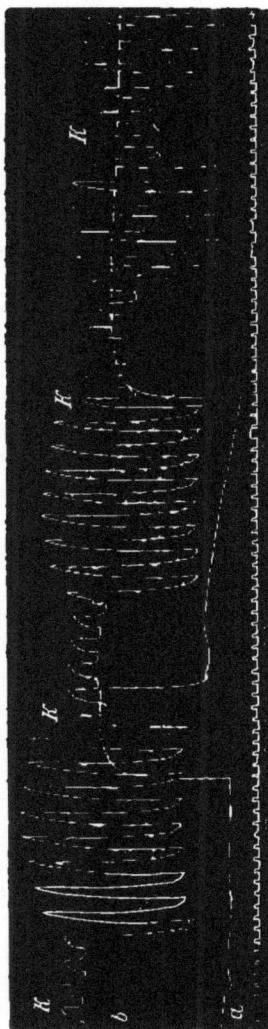

Fig. 43.—Artificial respiration after division of the medulla oblongata above the respiratory centre and the vagi in the neck. Respiratory spasms: a, without artificial respiration, b, during artificial respiration; at K, K inspiratory spasms.

that secondary products of the change of material are indirectly affected by the altered gases of the blood, and that these have a stimulating effect on respiration. On the other hand, after suspension of the insufflations, the previously intermittent conditions of spasm reappear tolerably suddenly, long before the blood has assumed the state of asphyxia. It is not explained by the gas-saturation hypothesis why it is more difficult to obtain apnœa after division of the vagi than when they are intact. According to the discoveries just described, the proposition does not seem improbable that the vagi (possessing tonus) not only facilitate the liberation of the shocks stored up in the centre, *but also facilitate the removal of the stimulating materials.*

I will not attempt to extend this hypothesis further, nor to attribute to the vagus a dilating action on the blood-vessels in the region of the *medulla oblongata.* For this I have no experiments to support my statements. However, the vaso-dilator (depressor) action of the centripetal vagus fibres has been proved by the famous discovery of Cyon and Ludwig.

Now comes the question whether apnœa is a normal or an abnormal condition. When the impulse for the liberation of respiratory movements is not sufficient it may have its cause in the fact that the excitability of the centre has diminished and that the shocks have sunk below their minimum power of stimulation. In fact, one sees in many cases, in which clearly the excitability of the brain is weakened, a condition taking place analogous to apnœa, as in coma and in poisoning by means of narcotics. I am inclined also to place the apnœa of the fœtus in this category. The same blood which rules the change of material in the respiratory centre of the mother also flows through the central nervous organs of the fœtus. Why is the respiratory centre of the first stimulated whilst that of the other remains at rest? But if we assume that the respiratory centre in the fœtus is less excitable, then one might ask why is it active in the new-born child? Certainly by the act of birth itself, through the original asphyxia, and through eventual peripheral stimulations, impulses to respiration are given; but why does the new-born child, by means of respiration just set into activity, not make itself immediately apnœic, and put itself into a condition corresponding to the Cheyne-Stokes' phenomenon? A suggestion may be found regarding this rather hazy question in the condition of the frog's heart, which has been thoroughly studied. The ventricle of the heart of the frog, as is known, immediately stops its pulsations when a ligature is applied below the auriculoventricular groove, so as to cut off the ventricle from the upper cavities. Each single shock coming to it liberates only one contraction. But when the ventricle of the heart, filled with blood or serum, is separated from the auricle for a long time, then it begins, as Merunowicz has shown, to beat automatically, and the beat no longer stops when blood containing oxygen is passed through it. It therefore appears that the nerve centres once stimulated for even a short time do not readily lose the effect of the stimulating impulse.

RESULTS.

The results obtained by this investigation may be thus expressed:—

1. In the *medulla oblongata* there lie in close connection with the roots of the vagus, the centres for respiration: a centre of inspiration and a more difficultly excited centre of expiration. During normal respiration the centre of inspiration alone is active, while the centre of expiration becomes active only under exceptional circumstances. It is an auxiliary centre of respiration.

2. There are no centres of respiration situated higher in the cerebro-spinal axis. All phenomena which have suggested the existence of such a centre can easily be explained as centripetal stimulations of the *medulla oblongata*, which, by reflex action, act on the respiratory centre in the fourth ventricle. The manner in which head-dyspnœa takes place and disappears is a fact against the existence of higher situated centres.

3. In the cervical part of the spinal cord there are only the central tracts of respiration; special centres for the liberation of respiration do not exist there.

4. The respiratory centre in the *medulla oblongata* not only acts automatically, but can be excited by reflex action.

5. The automatically active centre can only liberate respiratory spasms, not regular rhythmic respiratory movements.

6. Normal rhythmic respiration is a reflex act, mainly liberated by the vagi, which prevent the gathering tension in the centre becoming too great and convert the inherent stimulations of the respiratory centre into regular respiratory movements. The action is that of a " discharger."

7. The vagi constantly stimulate; they possess tonus, and are sufficient to serve as the only active regulators of respiration. During absolute rest of the animal organism they probably act alone.

8. Next to the vagi the upper brain tracts are of great importance for the liberation of regular rhythmic respiration. They are capable of replacing the non-activity of the vagi, just as the vagi may compensate the non-activity of the upper tracts. During sleep, during hibernation, or after certain narcotics the reflexes from the upper tracts on the respiratory centre often remain active.

9. If a portion of the upper tracts remains non-active while the vagi are still active, then periodic respiration may take place (Cheyne-Stokes' respiration).

10. For the accomplishment of periodic respiration a change in the excitability of the respiratory centre itself is not necessary. The latter can be excited during periodic respiration at the time of the periods as well as during the pause, equally powerfully by means of weak shocks of the same strength.

11. The sensory nerves of the skin cannot take the place of the brain tracts or of the vagi.

12. While the cutaneous sensory nerves in animals which are intact exercise very gentle, perhaps almost no, influence on the respiratory centre, this influence increases greatly after the brain tracts become non-active. In these circumstances, skin reflexes are capable of liberating complete series of respiratory movements. The cutaneous nerves possess no tonus.

13. The centripetal nerves, which have an inhibitory action on the liberation of respiratory movement, as, for instance, the *trigemini, laryngei superiores*, and *glossopharyngei* possess no tonus, and must be regarded as auxiliary nerves of respiration; during normal liberation of respiration they are not in activity.

14. The *trigemini* and *laryngei superiores* (as well as olfactory

and splanchnic nerves) on stimulation cause slow respiration and at last stop it altogether in a position of expiration, while the *glossopharyngei* have no influence on the frequency of respiration. The action of the latter is always an absolute inhibition of respiration, and this in every phase of the same in which the stimulation becomes active.

15. A single electric shock cannot of itself set the respiratory centre into activity, but only when its action is strengthened by other (intracentral chemical) stimulations.

16. The shortest irritation of the respiratory centre that can be caused sends to the phrenics four active, simple shocks, at intervals of about $\frac{1}{20}$ sec.

17. The normal stimulation of the respiratory centre does not depend on the blood; neither from deficiency of oxygen nor from too large a quantity of CO_2 in the blood. Animals continue to breathe without a circulation, and after bleeding, for a considerable time.

18. The Hering-Breuer mechanical respiratory theory is just as far from being correct, because after opening of the pleural sacs, and even after extirpation of the lungs, rhythmic respiration continues.

19. The active stimulating matters of the respiratory centre itself are probably of a similar nature to those which stimulate the isolated heart; perhaps the products of decomposition of intercellular fluids.

20. Apnœa has nothing to do with the gas contained in the blood, but depends probably on the removal of the stimulations stored in the centre, by means of the vagi in a state of tonus. Therefore after section of the vagi the production of apnœa is difficult, and it lasts only for a short time.

21. During apnœa one can neither by means of direct stimulation of the *medulla oblongata* nor of the central vagi liberate respiratory movements.

22. The respiratory centre of the fœtus during respiratory rest in the uterus of the mother is in a condition of much lower excitability than after birth.

23. When central or peripheral stimulations during the act of birth have once caused respiration, then the continuously-stimulated respiratory centre does not readily lose the stimulating impulse, and respiration goes on in a regular rhythmic manner.

24. Different phenomena, as the common-salt dyspnœa and the cold dyspnœa (heat dyspnœa?) make it probable that the phrenics carry stimulations direct from the surface of the brain, from a motor centre in the same, evading the respiratory centre in their passage downwards.

25. The presence of sensory fibres in the phrenics as well as the extremely delicate muscular sensation referred to the diaphragm support this view.

26. Rabbits, from the age of 4–5 months, after division of the phrenics may live; younger ones die from insufficiency of air in consequence of incomplete expansion of the thorax.

APPENDIX I.

ON THE PROPAGATION OF STIMULATION AND INHIBITION FROM THE CENTRE OF DEGLUTITION TO THE CENTRE OF RESPIRATION.

In my investigations "on the movements of respiration and their innervation in the rabbit," I came to the conclusions that the movement of the diaphragm, which takes place simultaneously with the movement of swallowing, the so-called "respiration of deglutition," is (1) not a true respiratory movement; and (2) that it depends on a passive movement of the diaphragm—a movement caused by the wave of contraction which follows swallowing. These observations were deduced from experiments on animals whose upper brain tracts had been removed. A closer examination of this question on intact animals proved to me, as I shall briefly explain in the following pages, that my second conclusion was wrong, and that the "respiration of deglutition" is in reality an active occurrence. On the other hand, my first conclusion was confirmed, viz.: that the movement of the diaphragm, originating with the movement of deglutition, could not be regarded as a movement of respiration. Undoubtedly in the respiration of deglutition we have to do with an irradiation from the centre of swallowing to the centre of respiration, but the mechanism is much more complicated than investigators, especially Steiner, have represented. Investigations carried on up till now have led to no definite conclusion regarding the nature of the "respiration of swallowing," or of its importance in the mechanism of deglutition, for two reasons: first, the movements of swallowing and of the respiration of swallowing produced by stimulating the laryngeal nerve were almost exclusively studied, whilst the study of the changes which the normal "period of respiration" sustained when swallowing was produced in a natural way was quite neglected. But the exact cause of these changes was much more important for understanding the "respiration of swallowing" and its influence on the act of swallowing, than the study of the respiration of swallowing itself, as it takes place under abnormal circumstances, such as are caused by stimulation of the superior laryngeal nerve. Secondly, the graphic method of registration in use up till now was insufficient and could not give a correct picture of the changes in the normal duration of respiration caused by swallowing, as these changes were too small and took place too quickly. The diaphragm-lever constructed by Kronecker and myself is, on the other hand, a very light, easily handled, and excellent instrument for giving a true representation of very small changes of movement of the diaphragm.

Rosenthal had seen the small diaphragm contractions during stimu-lation of the superior laryngeal nerves which interrupted the curve of the respiratory movement, but had mistaken their nature and cause. Then came Bidder and Blumberg, who declared these movements to be dependent on the action of swallowing, but considered them passive and caused by the impulse of the trachea and œsophagus on the lungs and diaphragm. The active nature of the occurrence was first pointed out by Waller and Prévost, in so far as they showed that these move-ments were isochronous with the first action of swallowing (at the same time as the elevation of the larynx), that they continued after the trachea and œsophagus had been cut across, and that they dis-appeared when the phrenic nerve (as they thought) had been cut. Arloing studied the changes on the respiratory mechanism during the accomplishment of natural swallowing, and found that in mammals at the time of the first stage of swallowing there was always an inspira-tory movement of the diaphragm, during which the glottis became larger. In birds a movement of expiration took place, and he founded thereon his theory of the action of swallowing. Arloing further drew attention to the fact (which Longet had already observed) that with the associated movements of swallowing, respiration became slower and sometimes was quite interrupted; thus the respirations of swallowing in the first case were marked on the respiratory curve as small jerk-like elevations. Before Arloing, Chauveau and Toussaint described an inspiratory movement of the thorax during the act of rumination, by which they tried to explain the mechanism of rumination. Then followed Kronecker and Meltzer, who again demonstrated the active nature of respiratory swallowing, and made it an accepted fact, and further showed the central connection between the centre of swallow-ing and the centre of respiration (as well as with other centres of the *medulla oblongata*). Meltzer regarded the respiration of swallowing as an expiratory movement of the thorax in which the diaphragm played a passive part and was moved in an opposite, *i.e.* inspiratory, sense; but still in 1% of the cases there was only an expiratory movement of the diaphragm. He came to such a forced conclusion probably because he felt keenly the contradiction and could not explain, that the same process of irradiation from the centre of swallowing to the centre of respiration should diminish the necessity for respiratory movements (as Meltzer himself discovered) and at the same time cause an apparent diaphragmatic inspiratory movement.

All observers agree that swallowing-respiration is quite short, and occurs in a jerky manner; the most of them also agree that it is very limited, as if it had been checked. A year after Meltzer's dissertation regarding the *Centre of Swallowing, its Irradiations, and its General Importance,* there appeared a treatise by Steiner which referred to the same subject, without bringing forward any really new views or showing any new facts which could clear away the mist with regard to the importance of the "respiration of swallowing" in the mechanism of swallowing. Still, Steiner's work is a lucid contribution, in so far as on the one hand it confirms the facts discovered by Waller and Prévost (as had also been done by Meltzer), that the "respiration of swallowing" is an active occurrence; and on the other hand, it led to similar considerations regarding irradiations of a central nature, as Meltzer had already pointed out. But it is to be regretted that Steiner never mentions the important works of Waller and Prévost as well as that of Meltzer, especially when he speaks of the "newly-discovered relations in which the centre of swallowing and the centre of respiration stand to one another." Steiner has added two new observations regarding the "respiration of swallowing:" (1) That it continues notwithstanding the occurrence of apnœa; and (2) that under certain circumstances (strong poisoning by morphia) an animal when swallowing always makes a slight movement of the diaphragm, not only as long as it breathes, but also after it has ceased to breathe, provided the respiratory centre can be stimulated by reflex action.

It is evident that all the discoveries made up to the present time do not solve the difficulty regarding the importance of the so-called respiration of swallowing. If simultaneously with each deglutition a respiratory movement took place, no matter how often the animal might successively swallow, the danger for the individual would be exceptionally great, and the "respiration of swallowing" would lead to a pneumonia. Further, a swallowing movement of respiration appears in contradiction to the well-known fact that respiration itself is suspended during swallowing; it is also at variance with the circumstance discovered by Meltzer, namely, that even the necessity for respiration is reduced during swallowing, so that by means of repeated acts of deglutition the respiratory pause can be lengthened by a number of seconds. If we add to this the observations which I made at an earlier period, namely, that in animals from which the cerebrum had been removed the respiration of swallowing produced by stimulating the superior laryngeal did not correspond, in form or duration, with those respiratory movements which we saw normally in animals operated on in the same manner, or which could be liberated either by reflex action from the vagus, or from any point on the skin of the body, or by direct stimulation of the *medulla oblongata.* The respirations produced by the last-mentioned method were always similar to normal respirations, while the movement of the diaphragm going on with deglutition was flat and short, as if arrested. Further, although in animals, treated as above mentioned, the vagi were divided, so that the respiratory spasms which I have described took place, distinct respirations of deglutition followed each stimulation of the laryngeal nerves or the posterior surface of the velum. On the other hand, during the respiratory spasms stimulations of the laryngeal nerves could not produce cessation of respiration in the position of expiration. All that could be obtained was only rhythmic interruption of the respiratory spasms with somewhat lengthened expiration. It was therefore the more remarkable that on stimulation of the superior laryngeal nerve under these circumstances the respiratory centre should answer the apparently weak and short stimulation, going from the centre of deglutition with a normal respiratory movement, which took place quite independent of the respiratory spasms. Stimulations of the skin, which in animals deprived of the cerebrum are very effective and always liberate complete respiratory movements, were no longer able to do this after division of the vagi; here all that was produced was an expiration during an inspiration, and an inspiration during an expiration. If, on the other hand, we caused these animals to swallow by simple mechanical stimulation of the posterior surface of the palate, then the respiration of swallowing showed itself in an even more marked rising and falling movement of the diaphragm.

In the following experiments I sought to reconcile these contradictions, and to settle the true nature of the respiration of swallowing and its importance in the mechanism of the act of swallowing. For this end, I shall enter minutely into the relation between the centre of deglutition and the centre of respiration, especially with regard to the position of the centre of deglutition.

My experiments, which were all made on rabbits, gave the following results:—

(1.) *The respiration of swallowing commences only about* 0·02 *—0·03 secs. later than the contraction of the mylo-hyoid muscles,* and this contraction (as Meltzer proved) represents the first stage of deglutition, while the contraction of the mylo-hyoid muscles precedes the elevation of larynx by 0·07″. Further, the elevation of the larynx exceeded in duration the contraction of the mylo-hyoid muscles by a considerable period of time (0·1 sec.). On the other hand, the "respiration of swallowing" was completed before the mylo-hyoid muscles had ceased to contract. This last-mentioned difference in time varies according as the deglutition is produced from the velum or from the superior laryngeal nerve; in the latter case the respiration of swallowing was shorter (0·3 — 0·4 sec.), in the first case longer (0·4 — 0·5 sec.). The duration of contraction of the mylo-hyoid muscles also changes, according to the intensity of deglutition, between 0·44 — 0·62 sec.; the oftener the action of deglutition is accomplished in rapid succession the longer it appears in duration. In the same way the duration of elevation of the larynx varied (0·48—0·65 sec.). The arrangement of the experiment which made it possible to obtain these data was very simple. The mylo-hyoids and larynx were, by means of hooks and threads, attached to the levers of two of Marey's receiving tambours, from which the movement was conveyed to a recording tambour which also, from the same ordinate, marked with the diaphragm lever on the drum of the Baltzer kymograph even the quickest different movements, while a tuning-fork, tuned to vibrate 100 times per second, acted as the time recorder.

(2.) *At the time of the respiration of swallowing the glottis is opened.* I inserted a T canula into the trachea of the rabbit and closed the limb leading to the outside, so that the animal breathed in the normal way. If I then stimulated the posterior surface of the velum with a quill pen the animal performed the actions of deglutition at each stimulation, and the "respirations of swallowing" were marked on the diaphragm curve. If, on the other hand, the uninjured superior laryngeal nerve was stimulated, then the animal performed deglutition, but the "respiration of swallowing" was no longer present, because the laryngeals had simultaneously shut the glottis. If now I opened the side tube of the T canula, so that the air entered unhindered, then on stimulation of the superior laryngeal nerves the respirations of swallowing again appeared. The same result was obtained when I divided the laryngeal nerve at the periphery before closing the T canula on the outside. That is, the "respirations of swallowing" on the diaphragm only become noticeable when air can enter the trachea; therefore under normal conditions during the "respiration of swallowing" the glottis must be open.

(3.) *The "respirations of swallowing" remain after division of the phrenic nerves.* In rabbits after division of the phrenics, the thoracic respiration, which was before latent, now appears, and the "respirations of swallowing," whether the deglutition is produced by stimulation of the superior laryngeals or by stimulation of the posterior surface of the velum, are recorded as very small movements of the thorax in which the diaphragm also participates, but only in a passive manner, and in quite the opposite sense from active diaphragmatic inspiration and expiration. It is accordingly through the conduction of the deglutition stimulation to the respiratory centre, after division of the phrenics, that the muscles of subsidiary respiration are set into action and that "respiration of swallowing" remains.

(4.) *If the first stage of deglutition—the contraction of the mylo-hyoids—does not take place, this has no effect on the occurrence or form of the "respiration of swallowing."* As was already mentioned by Wasselieff, by separation of the *medulla oblongata* at the level of the *tubercula acustica* the connections between the centre of deglutition and the motor tracts of the *trigeminus* are divided, then it is impossible to liberate any movement of deglutition through the sensory ends of the *trigeminus* in the velum, as is plainly apparent. On the other hand, on stimulation of the laryngeals, in addition to stoppage of respiration, one sees the larynx elevated and depressed as usual, the thorax and œsophagus contract in the normal manner, while the mylo-hyoid muscles are quite inactive and remain so during the first stage of swallowing. But on each occasion, before elevation of the larynx, the respirations of swallowing appear in the usual manner on the curve line of respiratory cessation.

(5.) *During apnœa the respiration of swallowing, as Steiner observed, remains unchanged.* On a previous occasion, I was able to prove that in animals from which the cerebrum had been removed, reflexes from the skin remain without effect during apnœa; that, further, neither by means of direct stimulation of the *medulla oblongata* nor of the central roots of the vagus by single shocks, can respiratory movements be liberated. In these circumstances, successive shocks give very flat and slow movements of the diaphragm, which do not resemble the normal respiratory movements. We now see that the irradiation from the centre of deglutition to the centre of respiration in a condition of apnœa is capable of liberating an apparently small respiratory movement; the respiratory centre is accordingly not inexcitable during apnœa; on the contrary, a very slight shock can stimulate it by reflex action. I have often observed cases in which, during apnœa, the diaphragm came to a position of equilibrium in inspiration. If at this time a deglutition movement was liberated by irritating the velum, then the diaphragm, on the respiration of swallowing (generally after a short inspiratory movement), takes the position of expiration, from which it only returns gradually to the position of inspiration in apnœa. If the laryngeals are stimulated during "inspiratory apnœa," the diaphragm does not shift into a position of expiration, but retains its original position, and the progress of the "respiration of swallowing" is in no wise different from that caused by liberation of deglutition on stimulation of the velum during inspiratory apnœa.

(6.) *When the centre of respiration is removed the respiration of swallowing does not take place.* I have often scooped out the gray substance of the *ala cinerea* by means of a sharp, fine gouge. Respiration immediately ceased and artificial ventilation had to be set in action. During suspension of the latter, the posterior surface of the velum as well as the laryngeal nerves were stimulated, the first mechanically, the latter by means of electricity. The diaphragm-lever invariably recorded a straight line, but the animal performed deglutition in the normal way, irrespective of the manner in which the deglutition had been produced. From this the following was inferred:

(7.) *The centre of swallowing can act quite independent of the respiratory centre, and does so even after the latter has been destroyed.* An animal can therefore swallow after it has ceased to breathe (Steiner also observed this after poisoning with morphia). It further follows from this experiment that the superior laryngeal nerve contains fibres which go direct to the centre of deglutition without previously passing through the respiratory centre and without being influenced in their action by the respiratory fibres of the superior laryngeal nerve.

I must mention here some observations made in connection with the mode of action of the stimulated superior laryngeal nerve and which show the specific activity of the latter. If in the usual way the laryngeals are stimulated during ordinary respiration, the diaphragm stands still in the position of expiration, a fact known since Rosenthal's time. An expiration generally does not take place, but it may occur after long-continued or very strong stimulation. The respiratory centre during laryngeal respiratory stoppage can be stimulated reflexly, as is shown by the respiration of swallowing. Suppose the brain substance be bored close above the apices of the *alæ cinereæ*, and a little outwards from them, till the instrument passes through the floor of the fourth ventricle; this operation is often successful without specially injuring the respiratory centre. According to the injury sustained by the centre, respiration either remains normal or it becomes (when the centre is much irritated) very frequent and small. When the respiratory centre is partially injured, respiration either becomes periodic or deep and slow and interrupted by long pauses (intermittent). If in any of these cases the superior laryngeal nerve is stimulated, it is impossible, either by the normal or by quick or slow respiration—no matter the degree of stimulation adopted—to cause stoppage of respiration in the ordinary position of expiration of the diaphragm; there is only a long expiratory spasm with contraction of the muscles of expiration which inserts itself either between the normal or frequent respirations, or when the respiration is slowed it directly precedes the respiration or follows it immediately. Further, in periodic respiration in animals from which the cerebrum has been removed, it is possible by stimulation of the superior laryngeal nerve to remove the periods and to cause continuous respiration; if the brain substance be removed in the above-mentioned way and the laryngeal be stimulated, on each occasion before a respiration, active expiratory movements appear in a very characteristic manner. A removal of brain substance at the same level of the *medulla oblongata*, but at another position—for instance, on both sides of the raphe—has not the same result, although respiration sometimes becomes just as frequent or may be slowed as in the previous operation. Here there appears, as under normal conditions, upon stimulation of the *laryngeus*, cessation of respiration with relaxation of the diaphragm. It would appear that the superior laryngeal nerve has a double action on the respiratory centre: at one time it liberates expiratory spasms, acting as a coughing nerve κατ' ἐκοχήν; and at another time it only removes the tonus of the respiratory centre. With ordinary electric stimulation of the laryngeals the first, and, as we must suppose, specific, action of the laryngeals does not take place, and the question now is how it comes to be paralysed. The removal of the part of the brain substance which only leaves the cough-exciting action of the laryngeal nerve, concerns mainly the region of the nucleus of the glossopharyngeus, and it is conceivable that generally with stimulation of the laryngeals the glossopharyngeus is also stimulated and the tone of the respiratory centre lowered. We must accordingly distinguish three different kinds of fibres in the superior laryngeal: (1) Fibres which go to the centre of deglutition and stimulate it; (2) Fibres which inhibit respiration—that is fibres which remove the tone of the respiratory centre; (3) Specific coughing fibres which stimulate the centre of expiration and liberate fits of coughing.

(8.) *The inhibition of natural respiration is the constant and essential character of the respiration of swallowing, and this inhibition generally, but not always, is preceded by a short inspiratory movement.* From the centre of deglutition to the centre of respiration there radiates not only the stimulation of deglutition, as previously

assumed, but specially the inhibition of deglutition. In the transmission of this latter to the respiratory centre we must observe the
arrangement which protects the organism from the otherwise threatening danger of pneumonia caused by inhaling alimentary particles
(Schluck pneumonia). A close study of the respiration of swallowing
as it presents itself during natural respiration lends support to this
view. If the velum in a rabbit is stimulated we liberate a number of
deglutitions after one another, and the accompanying respirations of
swallowing make themselves obvious, during inspiration as well as
during expiration, and also during the respiratory pause, on the curve
of the natural respirations, as small indentations and notches. When
the respiration of swallowing takes place exactly at the end of the
inspiration or at the beginning of the expiration, the respiratory
movement at the apex of the inspiration appears as if it were divided.
If deglutition is caused several times in succession, the natural respiration becomes lower and at the apex of the same there is again an
indentation and a notch, indicating that respiration breaks off exactly
with the respiration of swallowing; the impulse of the natural inspiration after the respiration of swallowing comes no longer into play.
If deglutition is caused still oftener, respiration stands still and the
respirations of swallowing are quite similar to those we obtained
during the stoppage of respiration on stimulation of the superior
laryngeal. If the respirations of swallowing be recorded during
natural respiration, simultaneously with the contraction curve of
the mylo-hyoid muscles during the quickest rate of movement of
the drum of the kymograph, so as to note the beginning of the
"respiration of swallowing,"—according to my previous experiments,
0·02–0·03 sec. after the beginning of the contraction of mylo-hyoid
muscle,—then one observes that during inspiration the inspiratory
movement of the "respiration of swallowing" coincides with the curve
of natural inspiration. Then respiration suddenly stands still and an
expiratory movement of the diaphragm takes place, which often lasts
as long as the ascent, and is interrupted by the impulse of the natural
inspiration before it reaches the abscissa. On the descending limb
during natural expiration a similar occurrence takes place, only here
the inspiratory movement of the respiration of swallowing is plainly
recognized as ascent, while the expiratory movement of the same is

continued in the natural expiration. The expiration then continues slower. Natural respiration becomes considerably lengthened, whether it be that the "respiration of swallowing" runs into inspiration or into expiration. There are cases in which the movement of inspiration of the "respiration of swallowing" does not take place, and there is only an expiratory movement of the diaphragm. This is sometimes seen in inspiratory apnœa, as already mentioned; further after large doses of morphia, and finally in connection with a very interesting occurrence which one can compare to the first act of drowning. If a rabbit, upon which tracheotomy has been performed, and which can obtain air through a canula in the lower part of the trachea, is allowed to have water flowing continuously into the pharynx and nostrils so that the water runs away by means of a fistula in the larynx or in the œsophagus, there appears, after about 15″, restlessness, during which the animal does not breathe, but performs the actions of deglutition several times, a cessation of respiration, lasting about 5 minutes, which is interrupted by a few respirations at very long intervals. At the same time the diaphragm gradually goes over into the position of complete inspiration. While at the beginning, as long as the diaphragm still stood in the position of expiration the respirations of swallowing were as usual represented by small elevations and depressions on the curve of respiratory stoppage; during inspiratory stoppage they had a purely expiratory character. Further, at the commencement of deglutition the diaphragm always became relaxed and only returned gradually to its former condition of tonus. On the other hand, there are conditions in which it is just the inspiration of the "respiration of swallowing" which is specially well marked and can even resemble the depth of a small respiratory movement. But these are always experimental and pathological cases, like after transverse division of the *medulla oblongata* above the respiratory centre, especially when at the same time the vagi were divided, or with Cheyne-Stokes' respiration, &c. The form of the "respiration of swallowing" resembles in so far a respiratory movement, as a short inspiratory movement generally precedes the relaxation of the diaphragm, caused by the inhibition of respiration. The former might, nevertheless, frequently elude observation; for instance, each time the "respiration of swallowing" falls into the inspiratory phase of natural respiration, when it was not followed by inhibition, after which the diaphragm goes over into expiration. With the centre of swallowing and the centre of respiration close together, and considering their intercentral connections, it is not extraordinary that not only the inhibition of swallowing but also the stimulation of swallowing is irradiated. It would be more remarkable if only the one occurrence were transmitted and the other not.

(9.) *The transmission of inhibition from the centre of deglutition to the centre of respiration takes place by means of the glossopharyngeal nerves.* We know, from the experiments of Kronecker and Meltzer, that with each liberation of deglutition the first stimulation of the centre of deglutition is followed by an inhibition of deglutition, which comes into effect between the mylo-hyoid and pharynx deglutition movements. These authors further showed that the glossopharyngeal nerve is the inhibitory nerve of deglutition, and that its stimulation prevents deglutition coming into effect. My earlier observations have shown that the same glossopharyngeal nerve is also an inhibitory nerve of respiration, and, indeed, of a peculiar sort. It contains considerable latent energy and inhibits respiration in all its stages during inspiration, expiration, and the respiratory pause. We see, further, that deglutition can take place in every stage of natural respiration, and the curves show plainly that thereby respiration is always interrupted, and that the inhibition of deglutition is over when the inspiratory stage of the respiration of swallowing has reached its maximum. Then follows a reduction of the tone of the respiratory centre, so that the diaphragm relaxes somewhat, before it again follows the shortly interrupted impulse of the respiratory tonus. There remains no doubt that it must be the glossopharyngeal which inhibits deglutition as well as respiration. If the centre of deglutition be stimulated, then the nucleus of the glossopharyngeus connected with it is also stimulated. Stimulation and inhibition irradiate simultaneously to the centre of respiration. As the period of latent stimulation of the stimulated glossopharyngeal is greater, the stimulation arrives earlier at the respiratory centre and liberates an inspiratory movement of the diaphragm. Then the inhibition immediately follows, interrupts the inspiration, and hinders the completion of normal respiration. When swallowing takes place several times in succession, the accumulated inhibitions become active in so far as they decrease the natural tone of respiration. On the other hand, the "respiration of swallowing," as it immediately follows the act of swallowing, cannot be suppressed by frequent deglutitions. The character of the respiration of swallowing only depends on the fact that the stimulation of the glossopharyngeus irradiates from the centre of deglutition to the centre of respiration, and so interrupts the weak impulse of the vagus. Accordingly the movement which results from the secondary stimulation of the respiratory centre has not the importance of a respiratory movement. It is providentially arranged that alimentary masses in the pharyngeal cavity are not aspirated during the act of swallowing through the effect of the "respiration of deglutition." If a true respiratory movement takes place, the animal chokes and must restore itself by coughing, while by this means the centre of deglutition comes into a state of increased stimulation and brings the body ejected into the right tract. In addition to the glossopharyngeus there exist two other nerves which have an inhibitory action on respiration, one of which always, the other sometimes, takes part in deglutition. Might not the *trigeminus* nerve or the superior laryngeal nerve cause inhibition of respiration during deglutition? Without reference to the fact that neither the *trigeminus* nor superior laryngeal are capable of inhibiting respiration in all its various stages, the superior laryngeal takes no part in the accomplishment of deglutition when the same is liberated from the velum, therefore is not capable of influencing respiration during deglutition. The *trigeminus* nerve, as I have already remarked, can be so divided by a transverse section of the *medulla oblongata* above the centre of respiration from its connections with the centre of deglutition, that not only from the velum can no deglutition be liberated, but also on stimulation of the superior laryngeal the first stage of swallowing does not take place. But the respirations of swallowing continue unchanged, and thus the irradiation of the inhibition of deglutition to the centre of respiration can only take place by way of the glossopharyngeals.

Steiner in his works on the centre of respiration made out a scheme in which a connection between the centre of deglutition and the centre of respiration is effected by an intercentral fibre, which conveys the stimulations from the first to the second centre. This view is not

sufficient to explain the respiration of swallowing. The centre of
deglutition, the nucleus of the glossopharyngeus, and the respiratory
centre, are all in communication by intercentral connection, and at
the liberation of every deglutition stimulation and inhibition are,
as already mentioned, simultaneously irradiated to the respiratory
centre. But why, if intercentral fibres run in both directions, does
a transmission from the respiratory centre to the centre of deglutition
not take place? According to Steiner's arrangement, it is not explained why it is that, although along with each deglutition a respiration takes place, a deglutition does not occur with each respiration.
We must seek the reason in the following.

(10.) *The respiratory centre acting steadily and rhythmically
possesses a tonus* by means of which stimulations flow steadily to the
nucleus of the glossopharyngeus as well as to the centre of deglutition,
and thus alternately counteract one another.

(11.) *The centre of deglutition on the other hand possesses no tonus.*
Stimulation of deglutition and inhibition of deglutition with every
liberation of deglutition are stimulated and conducted anew.

(12.) *The centre of deglutition in the medulla oblongata can be
destroyed without in any way injuring respiration.* If with a sharp
dagger-shaped pin or a fine auger the brain substance is removed at the
level of the apices of the *alæ cinereæ*, and a little towards the median
line, while the animal's head is placed at a right angle to its body,
and the instrument passes through the floor of the fourth ventricle,
a spot is reached the destruction of which stops deglutition. In these
circumstances the mylo-hyoid muscles, the larynx, the pharynx and
the œsophagus remain completely inactive. Respiration, on the other
hand, goes on quite undisturbed. Stimulation of the superior laryngeal
nerve has the same effect as usual, the diaphragm remains stationary
in the position of expiration; but the curve of respiratory cessation
drawing a straight line is not interrupted by the small respirations of
swallowing. The microscopic examination of such brain cuts, which
had been injured in the before-mentioned manner, showed, that in
fact one succeeded in destroying by the sharp pins single circumscribed masses of ganglionic cells lying close to the cell group of
the nucleus of the vagus. These cells are situated about the inner
portion of the *pedunculus cerebelli*, a little above the visible points of

the *alœ cinereœ*, a region in which Vejas has discovered ganglionic cells.

If the removal of brain substance be carried on at some distance from the above-described spot, then the act of deglutition remains more or less complete. Thus every degree of effect can be obtained: cases in which only the first stage of deglutition does not take place; cases in which deglutition—no matter how it be liberated—appears very seldom and with difficulty; other cases in which deglutition is liberated from the velum alone; and lastly, cases in which the laryngeal deglutitions alone do not take place.

(13.) *By direct electric stimulation of the medulla oblongata deglutitions cannot be produced.* In vain have I attempted by means of direct electric stimulation of the *medulla oblongata* to obtain a liberation of deglutition. When one considers that by such stimulation it is difficult or even impossible to stimulate the centre of deglutition alone without at the same time touching the nucleus of the glossopharyngeus, and in this way inhibiting deglutition as soon as it is liberated, it will be seen that this method cannot give useful results.

These results may be thus summed up:—(1) The centre of deglutition, the glossopharyngeus nucleus, and the respiratory centre stand in the closest connection with one another. But deglutition takes place in a very different manner from respiration. The respiratory centre is automatically active, and the constant stimulation of the vagus-ends, having tonus, facilitates the liberation of inspiratory and expiratory movements only. In the respiratory centre continuous stimulations are converted into rhythmic movements, and it requires no inhibition to regulate normal respiration. We can alter the rhythm and depth of the respirations at will. (2) The centre of deglutition behaves in quite a different manner. Generally it is not stimulated, and it is only the stimulation of deglutition which from time to time sets it into action. But then we find a number of nerves which are always stimulated invariably in the same way and in the same order. The will has no influence on the action of swallowing, and once it has started it takes place without alteration in character. The stimulation of the centre of deglutition, proceeding from the periphery, is followed by the inhibition through the glossopharyngeal nerves before pharyngeal deglutition begins, and in this way a number of deglutitions can follow one another in quick succession. The fibres of deglutition of the vagus and glossopharyngeus thus stimulated irradiate the stimulation simultaneously to the centre of respiration. Following the stimulation of the fibres of deglutition of the vagus the respiratory centre answers with an inspiratory movement. The stimulation of the glossopharyngeal fibres immediately inhibits this. Thus the inspiratory movement appears so short, and it is only the prolonged period of latent stimulation of the glossopharyngeus that makes it visible at all. *Therefore the so-called " respiration of swallowing" has not the importance of a respiratory movement, but it represents a very important occurrence in the action of deglutition, namely, the inhibition of respiration.* This interruption of respiration during deglutition is the most effective protection the body possesses against the dangers of swallowing the wrong way.

The theory of the inhibitory action of the glossopharyngeals on respiration gains fresh support from the investigation. The results of this paper connect in a natural way the action of the inhibition of deglutition by those nerves, which, as Meltzer has proved, have also the power to remove the tonus of the vagus on the heart.

APPENDIX II.

In my work on the *Respiratory Movements in the Rabbit and their Innervation*, which embodied the results of experiments carried on for years, I was compelled to leave many questions unanswered, for which I needed further researches. The various new theories which I started naturally found many antagonists. Some of the objections, partly critical and partly founded on experiment, to my views I have already answered in a communication on the "Propagation of Stimulation and Inhibition from the Centre of Deglutition to the Centre of Respiration." Prof. H. Kronecker, in a paper entitled "Old and New Facts about the Respiratory Centre," has also cleared up various difficulties and misunderstandings.

I wish now to explain more fully the point which is most important and regarding which most difficulty has arisen, namely, the "Respiration of the Spinal Cord," for which, on the basis of new experiments, Langendorff has again broken a lance. I am also obliged to do this, because immediately after the publication of my work there appeared a paper by Wertheimer entitled, "Sur les Centres Respiratoires de la Moëlle Épinière," which, by an intelligent interpretation of experiments, seemed to corroborate the view that the centres for respiration are situated in the spinal cord, and that the *medulla oblongata* has only a regulating and moderating effect on the respiratory mechanism of the spinal cord.

Unlike the method of Langendorff, Wertheimer in his experiments used only old animals, and these were always dogs ranging in age from a few months to full-grown animals. Of the 200 animals experimented upon, only 56 (only about ¼) gave the expected results. Dogs from 3 – 6 months old or a little more seemed to be the most favourable subjects. Full-grown animals, however, lived for hours after division of the spinal cord, and during this time rhythmic respiration again went on.

None of the animals were put under the influence of strychnine. The *medulla spinalis* was divided transversely immediately below the *calamus scriptorius*, and directly, according to the age of the animal, artificial respiration, lasting $\frac{1}{2}$–5 hours, was put in operation until spinal respiration came into action. When this had been done and when the artificial respiration had been suspended, the animals breathed unaided for 25 – 30 – 45 minutes, even in cases where the abdominal muscles were separated from the thorax and the abdominal cavity and diaphragm had been laid open. The number of respirations in a minute varied from 50 to 90, 100 and even 130. The more frequent respiration was, the more superficial and irregular the contractions of the diaphragm became. If artificial respiration was renewed from time to time, respiratory movements might be recorded for hours.

Wertheimer explains that the stoppage of respiration immediately after division of the spinal cord is due to shock caused by the operation, which has an inhibitory action on the spinal centres of respiration, and in the same way that destruction of the *noeud vital* does not destroy a centre situated higher up, but stimulates the deep-lying centres so violently that they immediately suspend their power of action. When the *medulla spinalis* recovers from shock, the more distant expiratory centre begins to act sooner than the inspiratory centre, which is situated nearer the cut surface, and respiration has at first a purely active expiratory character. When division of the *medulla spinalis* was intentionally incomplete and spontaneous respiration had again appeared, the latter was not hindered on completion of the incision. *Therefore the new incision did not inhibit the centres of respiration.*

Wertheimer found further, that the activity of the spinal respiratory centres begins (already in opposition to the observations of Langendorff) during even very complete artificial respiration, and he concludes that it neither requires asphyxiated blood to stimulate the spinal centres to activity nor reflexes coming from centripetal nerves. When the centres once become active they continue to work automatically, for the reason that they are nourished by normal blood. From his experiments, Wertheimer argues that Brown-Séquard's theory that the respiratory movements depend on all the excito-motor parts of the cerebro-spinal axis, and on the gray substance which connects these parts with the respiratory muscles, is the only theory which explains all the facts. If, according to this theory, one assigns the name *centre* to all the regions which can alter the rhythm of respiratory movements, then it is possible to increase the centres to any number. But, after Wertheimer, the notion of centres must be narrowed; a centre must not only be stimulated by reflex action, but must also be automatically active. Therefore, one must consider as a centre the whole column of gray substance from which the respiratory nerves of the head and trunk arise; and this mass must not be considered as an anatomical centre only, but also as a functional centre of special groups of muscles. The spinal cord, says Wertheimer, is the immediate centre for the respiratory movements of the trunk; separated from the *medulla oblongata*, it continues to send rhythmic impulses to the thoracic and abdominal muscles, and its tonus and reflex activity survive. The removal of the higher centres, in fact, causes rhythmic activity of the spinal cord to become more distinct, and the result is an excess of activity. This excito-motor power of the *medulla spinalis* becomes very plain on exposing the respiratory muscles and observing their frequent and prolonged contractions. Wertheimer further states that the anterior and posterior extremities take part in these movements and follow the same rhythm. "This inclination of the *medulla spinalis* to liberate periodic contractions, after it has been divided and separated from the brain centres, also takes place under other circumstances, even when these movements do not exist under normal circumstances." Wertheimer declares, as already mentioned, that the centre in the *medulla oblongata* has only a regulating and modifying action on respiratory movements.

According to this, the theory of respiratory innervation appears simple indeed! You only require to divide the spinal cord of a dog below the *calamus scriptorius*, connect the respiratory passages of the dog for some hours with a pair of bellows, in order to see the real and pure respiratory movements taking place—movements of all the respiratory muscles occurring in disorder, and following each other in frantic haste, the inspiratory centres in conflict with the expiratory ones; now the one, now the other carrying the day; and in this conflict the blood has no influence. In such respiration there is neither dyspnœa nor apnœa, nor is respiration excited by the blood, but, when once set going, the action of the blood (however it may be composed) on respiration is always the same in the sense of causing movement. Moderation and regularity in this confused mechanism is brought about by the *medulla oblongata;* it forces the expiratory centre to inactivity; it increases the depth, and reduces the frequency, of respiratory movement; and according to the nature of the blood bathing it, it causes dyspnœa, or well-regulated respiration, or apnœa, &c.

But is it not strange that both the supporters of the spinal-cord respiratory centres, Langendorff and Wertheimer, in their publications, came to conclusions quite opposed to one another? While Langendorff describes spinal-cord respiration as deep, slow, and regular, just like respiration when the vagus had ceased to act, Wertheimer describes it as quite superficial, very frequent, and very irregular; again, Langendorff finds that artificial respiration causes the centres to pass into a state of apnœa, but Wertheimer is of the opposite opinion; Langendorff regards asphyxiated blood as the cause of the stimulation of the respiratory centres of the spinal cord, but Wertheimer finds that the blood has no influence on the accomplishment of respiration; Langendorff affirms a participation of the expiratory centre only when the inspiratory centre is exhausted, or the deficiency of oxygen in the blood has become very great, Wertheimer, on the other hand, is of opinion that the expiratory centre becomes active immediately after the *medulla spinalis* regains its excitability, and that it remains longer excitable than the inspiratory centre. These numerous and important differences in their observations are certainly not to be explained by the fact that the one investigator, Langendorff, used newly-born or quite young animals, preferably rabbits, but also cats and dogs, for his experiments; while the other, Wertheimer, used only older dogs. Further, when Langendorff, in his paper on the "Innervation of Respiratory Movements," says, "The coincidence with Wertheimer in the results of the chief experiments, and what to me seems most valuable, the coincidence in the conclusions drawn from the experiments, is so great that compared with them unimportant minor differences need not come into consideration," I think no reader will agree with him in this statement, but perhaps, on the contrary, will conclude from the contradictory results of experiments carried on under not very different conditions, that the inferences must somehow be wrong.

When we consider the facts which both experimenters observed, from which they have drawn similar conclusions, there are only two, viz.: (1) that after section of the *medulla spinalis* below the *calamus scriptorius* respiration ceases; and (2) by using proper means, and a short time after section, one may see not only reflexes returning to the respiratory muscles, but also spontaneous rhythmic contractions of all the muscles taking part in respiration, similar contractions taking place in the muscles of the extremities. From these facts, which no one will question, the authors conclude wrongly that the centres of respiration lie in the spinal cord, while the *medulla oblongata* only takes the part of a regulator. They also hold, that when respiration stops after section, it is not because the respiratory centre has been separated from the centrifugal respiratory nerves, but because by this division, the more deeply situated actual centres of respiration are thrown out of action in consequence of the shock produced on the *medulla spinalis* by the operation. This shock inflicts an injury on the nervous apparatus of the spinal cord, and a simultaneous stimulation of the fibres which then have an inhibitory action on it.

In my work on the innervation of respiration in the rabbit, I have described at length and exemplified by experiments and tracings, that the contractions of the respiratory muscles obtained by direct electric stimulation of the *medulla spinalis*, as well as the reflex and spontaneous contractions appearing in young animals under the influence of strychnine, after section of the *medulla oblongata*, are not normal respiratory movements, but spasms of the respiratory muscles.

The doubts raised by Langendorff regarding the proper interpretation of my results, as Kronecker has already shown, are not justified, and are founded on a misunderstanding. To settle this point, it would be necessary to repeat at length what I said before, but I do not feel inclined to do this. The reader can decide for himself. I have further brought forward evidence from the writings of others that after the death of man as well as of animals it was possible to obtain rhythmic contractions of different muscles even after they had been separated from the body. This is specially true with regard to the diaphragm.

Kronecker has also mentioned several interesting examples in his paper; and I drew attention to the fact that in decapitated animals (such as the so-called reflex frog) continuous peripheral stimulations produced rhythmic movements, and yet no one has considered all these cases as giving sufficient reason for concluding that in the spinal cord there are motor and spasmodic centres. In older animals also, not under the influence of strychnine, but in exceptional circumstances, such as when the blood had been reduced in temperature so much that the heart removed from the body continued to beat for a long time like the heart of a frog, I have seen a few spontaneous and reflex respiratory muscle spasms take place after division of the *medulla oblongata.* Langendorff is of opinion that these are unfavourable conditions, and he says further that when an experiment goes on for some time, while artificial respiration is continued, the reduction of the temperature of the animal experimented on exercises a very unfavourable influence. "The reduction of the temperature of blood in a mammal is always accompanied by an enormous diminution in function of the central nerves." Wertheimer, on the other hand, states: "Si nous avons pu obtenir chez eux (les mammifères adultes) les resultats, que nous venons de mentioner, c'est qu' instruit par les recherches antérieures, nous avons eu soin de prolonger l'insufflation pulmonaire jusqu'au retour de l'excitabilité médullaire." And this lasted, as already mentioned, ½ – 5 hours, according to the age of the animal; and yet after this time he obtained a "veritable anhélation du tronc." Animals ventilated by artificial respiration for such a length of time have without doubt become cold-blooded. Wertheimer does not mention the temperature of the animals used in his experiments, but I have taken the temperature; and although when I repeated Wertheimer's experiment the air conveyed to the animal was passed through warm water, yet after 3–4 hours of artificial respiration the temperature in the rectum of a full-grown dog was 27° C. and even lower. The repetition of Wertheimer's experiment has almost made my former view a certainty, that it is just this artificially-obtained cold-blooded condition of a vertebrate animal that constitutes the essential condition for stimulation of rhythmic muscular spasms. The younger the animals are, the sooner will this cooling take place; so that in such animals one only requires to produce artificial respiration for a short time in order to see the spasms which, as Langendorff has shown, can be brought on more quickly and strengthened by small doses of strychnine.[1]

At the beginning of last winter, and shortly after the appearance of Wertheimer's work, I repeated many of his experiments on fully grown animals (dogs and cats), and in all the animals I always obtained similar results. With a clean cut, the *medulla spinalis* was divided immediately below the *calamus scriptorius*. Respiration immediately ceased. Then artificial respiration was produced by means of Kronecker's respiratory apparatus, with this modification that when the air reached the animal it was warm. Kronecker's apparatus enables artificial ventilation to be kept up, regularly and uninterruptedly, for any length of time. If one observes an animal in which artificial respiration has been thus kept up for 3 – 4 hours, its temperature will be found to be greatly diminished (27° C. in the rectum). One observes in these circumstances a very characteristic series of phenomena. While artificial respiration goes on uninterruptedly, the whole animal appears in constant, tolerably frequent rhythmic movement. Anterior and posterior extremities bend and straighten simultaneously in a spasmodic manner. The vertebral column follows these movements; with every bend the column is curved forwards, with every stretch it becomes straight. In the same way, the muscles of the neck and shoulders contract alternately. *The thorax is moved inwards and downwards with the same rhythm as the limbs*, and with apparently great force, and then jerks back into its position of rest. The movements resemble those of asphyxiated men when, by Sylvester's method of artificial respiration, the thorax is rhythmically compressed and the pressure is removed. The muscles of the abdomen in the same time contract spasmodically, and the air escapes from the trachea with a noise. The whole process represents rhythmic expiratory spasms of the thorax. I have never seen an actual raising of the ribs. The thorax during inspiration never went

---

[1] Since this was written, a new pamphlet, by Wertheimer, has appeared in which he states that by cooling the animal artificially down to 20° and 18° C., the spinal cord showed scarcely any signs of "shock" after division of the *medulla oblongata*, but that on the contrary, this process seemed to favour the return of spinal respiration. He observed in these cases, 5 to 15 minutes after *sectio bulbi*, that low, very frequent, and irregular respiratory movements reappeared. Wertheimer also states that in his former experiments the temperature of the dogs did not sink below 35° – 36° C.

past the position of equilibrium. These movements of the animal always take place in the same rhythm, and are not isochronal with the rhythm of artificial ventilation. In my cases, when artificial respiration was slow, they were analogous to the normal respiration of the animal, more frequent than those of artificial respiration, because the respiratory pauses between the spasms became shorter. After artificial respiration has ceased these spasmodic movements continue unchanged, perhaps for about an hour. If the abdomen is then opened, one can see and feel the diaphragm moving downwards with each expiratory spasm of the thorax in the manner of an inspiration; if the ribs are then elevated so that they no longer have any effect on the diaphragm, then these movements of the diaphragm immediately cease. This movement is, therefore, purely passive, like that seen after injury to the phrenics. Sometimes, besides this passive motion of the diaphragm, there takes place about every half-minute or every minute a true inspiratory spasm. This is very slow, and occurs at distinct intervals, and during it the diaphragm descends and becomes flat, and even moves downwards. During the relaxation of the diaphragm, one can also see rhythmic quivering movements synchronous with the strong and slow movements of the cold-blooded heart, and caused by its pulsations. The diaphragm, which has lost its tone, follows these movements in a very remarkable manner, and one must be careful not to be misled, and think that we have here to do with "very superficial, irregular, and frequent respirations." At last one observes quiverings which only affect single fibres of the diaphragm, and which continue even after the muscle has been divided. If, as Wertheimer did, all these different movements are registered by means of a tambour connected with the trachea of the animal, it is impossible to differentiate these different movements of the thorax and diaphragm, and of course one will arrive at wrong conclusions.

All the active movements which one observes in these vertebrate animals are so markedly spasmodic, and so different from the normal type, that it seems incomprehensible how they could be considered as the expression of normal respiratory movements. When the respiratory centre in the fourth ventricle was not destroyed, head-respiration remained in action during the whole time artificial respiration was carried on, and also continued for a long time after ventilation had ceased—much longer, indeed, than in warm-blooded animals—a sign that the *medulla oblongata* was greatly over-stimulated. The nostrils dilated and contracted, the mouth opened and closed with great force; but the movements of the head were independent of those of the trunk, and one might imagine that two different animals were under observation. This is the result obtained by keeping up artificial respiration for some hours in full-grown dogs and cats in which the *medulla spinalis* had been divided in the neck. Now it is evident that these muscle spasms do not all take place simultaneously and perhaps not always in the same order, and that by modifying the experiments in single cases and at certain times, one may observe the movements in the abdomen, diaphragm, and thorax alone; but it is also evident that both the movements observed during continued artificial respiration and those which take place at the commencement of artificial respiration are spasmodic. The character of the movements is not altered, but through time all the muscles of the trunk and the muscles of the limbs take part in the movement as a whole, just as the stimulation of the spinal cord increases. In my cases, the expiratory muscles of the thorax were stronger than the others. These, in combination with the abdominal muscles, prevented any heaving of the thorax taking place. At all events, the experiments of Wertheimer, as well as those of Langendorff, give no reason for concluding that there are respiratory centres in the spinal cord. "The tendency of the *medulla spinalis*," says Wertheimer, with regard to rhythmic movement of the limbs, "is to liberate periodic contractions," after it has been divided, and, therefore, when isolated from the brain centres; this also takes place under other circumstances, *even when these movements do not exist under normal circumstances.* Now, what is true of the muscles of the limbs is also true of the muscles of the thorax.

After an unprejudiced revisal of Wertheimer's experiments, I must repeat what I have already stated. "In the spinal cord there are no special centres for respiration. The single or rhythmic so-called respiratory movements emanating from the spinal cord in new-born animals, or animals in which the blood has been made cold, are not normal respirations, but tetanic conditions of the respiratory muscles; and very often other muscles take part. All such phenomena can easily be explained by facts known long ago."

K

Of course, with these conclusions, the assertion of Langendorff and Wertheimer, that the cessation of respiration after *sectio bulbi* is a result of the shock action of the *medulla spinalis* by which the respiratory centres in the spinal cord are inhibited and also injured, falls to the ground. When I undertake, nevertheless, to test this opinion directly, and to show that it cannot be sustained, I only do so to remove the last doubt of those who are still inclined, along with Langendorff and Wertheimer, to consider as actual respiratory movements those I have observed and explained as respiratory muscle-spasms.

With regard to the effect of shock after *sectio bulbi*, I have already explained that injury to the spinal cord either does not take place or very soon passes away, and while the reflexes again become normal, respiration does not return. Nor does shock injure, as Kronecker has shown, the centre of respiration in the *medulla oblongata*, either when the plane of section is above the origin of the vagus or when it is carried below it. Recently I performed the operation on marmots deeply asleep. These animals during hibernation behave in many respects like reflex animals. When the tail or the paws are pinched, then they move these, or they move corresponding limbs simultaneously, but with each movement respiration is immediately altered. Either a respiration is produced, or the respiration occurring at the time is deepened. The normal respiration of these animals during hibernation is very slow, a respiration occurring every half-minute or every minute, and very superficial, so that they are only noticed when very carefully looked for. When in these animals the membrane is opened, and the spinal cord in the cervical region divided close below the *calamus scriptorius* with a sharp cut, then no convulsion is observed. The reflexes continue to act, only respiration immediately ceases, and cannot be restored. The reflexes, indeed, after a few minutes become strangely marked. If you stimulate the tail, first the one and then the other hinder paw is moved; if you stimulate any of the paws, the corresponding paw on the other side is moved; when the hind limbs are strongly stimulated, the front legs are moved—and *vice versa*. No matter how often and how strongly you stimulate, a reflex movement of the respiratory muscles never takes place. The heart continues to beat for a long time, even for upwards of an hour.

Any notion of shock of the spinal cord after *sectio bulbi* in the marmot is therefore out of the question; and still, in these animals, after the operation, respiration immediately ceases.

Langendorff has also objected to the experiments in which I have shown that the trigeminal nerves are only subsidiary nerves of respiration, and that when they are divided the type of respiration is not altered; he also objects to my view that the vagi are not inhibitory nerves. He says that other observers hold a different opinion, and that the conditions of my experiments were made to suit myself, and only held good in such special conditions.

No doubt my opinion as to the function of the vagi is in many and important respects new, but I think it may be traced to improved methods of experimentation. I have minutely explained that it is necessary to extirpate the cerebrum in order to demonstrate the actions of the vagi on the respiratory centre, because the vagi contain fibres which cause pain, and may also convey stimulations to the *trigemini* and to other tracts which have an inhibitory action on respiration. For this reason, I divided the *medulla oblongata* transversely above the origin of the vagi, and showed that then respiration goes on quite normally, regularly, and automatically, and is not disturbed by any reflexes from the brain. In such animals I investigated the activity of the vagi, and found that they possessed their tone and were capable of liberating inspiration as well as expiration, but that they never caused active expiration. I also made out that the centripetal tracts of the spinal cord, as well as such nerves as the laryngeal and glossopharyngeal, possessed no tone, and that, when divided, they did not alter respiration in any way. In this manner I found that the respiratory centre in the fourth ventricle, separated from its centripetal nerves, was automatically active, but this automaticity only consisted in that it liberated respiratory spasms, but no regular rhythmic respiratory movements. These rhythmic respiratory movements are produced by the vagus, which prevents the tension accumulating in the centre in an unnatural manner, and converts the inherent stimulations of the respiratory centre into regular respiratory movements. For these reasons, and on the suggestion of Professor Kronecker, I called the vagus nerve a *liberator*, and denied that this nerve had any *inhibitory* action on respiration. Even those, however, who ascribe an inhibitory action to the vagus will not accept the view that in the marmot *sectio bulbi* inhibits the centre while it leaves the reflexes in their normal condition. If spinal-cord respiration existed, it would be specially apparent during hibernation of the marmot, because these animals behave like reflex animals. A further proof that it is not the shock which inhibits and injures respiration after *sectio bulbi* is offered by the experiments made by Schiff and Vulpian against the view of Brown-Séquard, namely, partial division of the spinal cord. From a report of some experiments conducted by Schiff in the Physiological Laboratory of Florence, entitled, " With regard to the Influence of the *Medulla Oblongata* on Respiration," I have taken the following details. Schiff says: In opposition to the doubts recently brought forward by Brown-Séquard, it has been lately confirmed, and by means of the graphic method made certain, that at the level of the first cervical nerves, the lateral fibres of the *medulla oblongata* convey the stimulation to the respiratory movements of the corresponding side of the body. Division of the lateral strands stops respiration on one side of the body in so far as it is not influenced passively by the other side or elevated by the abdomen. These passive movements are partly directed just in the opposite direction to the normal movements. On the occasion of the Medical Congress, a greyhound was shown in which, as had been done successfully on a former occasion, the lateral strands of the *medulla oblongata* had been divided at the lower border of the occipital foramen in such a manner that the movement of all four extremities had so far recovered in a few days that the animal nimbly ran about the room and could perform all ordinary movements. For six weeks, respiration was confined to one side, and after the animal had been killed by the inhalation of ether, and the abdomen been opened before the respiration had completely ceased, the absolutely one-sided respiratory movement of the diaphragm could be directly verified. When Brown-Séquard maintained that in rabbits, after mere injury to the lateral strands, he observed that respiratory movements were even increased, this probably arose from an incomplete injury which had a stimulating effect. In rabbits, Schiff attached a tube firmly to the trachea, which carried the inspired air through a Ludwig-Müller's double-ventilator into a graduated tube. After nine respirations, the volume of the displaced water was measured; then the *medulla ob-*

longata was laid bare, and the experiment repeated. As a general rule, the volume of respiration was not altered, but after division of either the half of the cord or only the lateral strands, the volume of the respiration was diminished by one-third.

On the 21st March, 1887, I opened the *membrana obturatoria* in a strong full-grown male cat, still young, and under the influence of morphia, and taking strict antiseptic precautions, I divided the whole right side of the spinal cord by means of a rapid incision with a cataract knife. The animal did not lose a drop of blood. On the left side respiration continued without interruption, while on the right side it immediately stopped. The right side of the body was completely paralysed. The wound was carefully stitched, and the animal put into its box. It immediately lay down on the right side in order to fully expand the thorax. When laid on the left side, it had the greatest difficulty in breathing, and immediately turned itself on to the other side. During the first few days the animal was very dull, and ate nothing; after the sixth day it became more lively, and took nourishment. On the paralysed side sensibility and motion were gradually but very slightly restored. On observing the thorax, one immediately noticed the want of any congruence between the movements of the two sides; while the left side of the thorax dilated, the right side moved inwards; and while the left returned from inspiration and often went into the position of active expiration, the right side again expanded in order to assume the original position of equilibrium. On the right side, an *active* inspiration or expiration was never observed. As regards the abdomen, the inequality of the movements was not observed, because here the bowel held the coverings of the abdomen on both sides equally stretched. The incongruence of the movements in the thorax continued unchanged during the whole time of observation, and the movements were taken graphically by means of two of Marey's tambours attached to a bent steel spring and joined by means of two india-rubber tubes, equal in length, to two recording tambours of the same shape. From the same ordinate the recording tambours commenced to write on a rotating kymographic cylinder. When the point connected with the left side recorded an inspiration, that on the right side marked an expiration, and *vice versa*. On the left side, sometimes quick small inspirations pushed between which did not necessitate any

movement of the thorax on the right side. It was very difficult to fix
the tambours without causing them to press on the thorax; they
easily changed position, and therefore did not allow of a long series
of respirations being recorded from the same ordinate. On the other
hand, the eye at once observed that the one point went downwards as
soon as the other went upwards, and *vice versa.*

On the 10th day after the operation the cat was again put
under ether, fixed down, and the wound opened. Under the skin
there was a little pus. Muscles and connective tissue had already
healed and had to be again divided by the knife. When the *calamus
scriptorius* and the upper portion of the spinal cord in the cervical
region were laid open, one saw very plainly the fine line-like cicatrix
on the right side of the *medulla spinalis*. At this spot the substance
of the spinal cord appeared narrowed and contracted. In the same
manner as in the first operation, the left side of the spinal cord in the
cervical region was divided, 1–2 lines above the former incision, and
again without the loss of a drop of blood. Immediately respiration on
both sides stood still, while head-dyspnœa increased. Artificial respiration had to be started with great speed; it was interrupted after a
quarter of an hour; again started and reinterrupted after another
quarter of an hour; still no trace of respiration in the trunk could be
discovered; the diaphragm when laid bare displayed no sign of movement, while the head continued to breathe with no special signs of
dyspnœa. This observation corresponds in every detail with the results
of experiments made on dogs by Schiff. Notwithstanding a lapse of
10 days, so that the possibility of the action of shock could no longer
be considered, and notwithstanding that sensibility and motion on the
injured side had already partially returned, still respiration did not
return to the right side. When the right side of the cervical cord was
divided, the respiration of the cat did not stop for a moment. No one,
therefore, has a right to assume that division of the left side, quite
as cleanly and quickly carried out, would inhibit respiration on *both*
sides. Even Langendorff admits that the more expert one is, then the
seldomer does long-continued stoppage of respiration on both sides
take place after division of the cervical cord on one side. This observation means nothing else than this, that one must avoid bleeding and
tearing, contusion, &c., of the sound side. That was the reason why

I went a little higher with the second incision, up to a level which, on the side first operated on, lay above the spot where conduction was interrupted.   No bleeding took place.

Wertheimer has also concluded from his experiments that if, after incomplete division of the *medulla spinalis*, spontaneous respiration again appeared, it was not stopped by completing the incision.   Consequently this operation on the cat shows unquestionably that respiration after *sectio bulbi* is not interrupted because the incision inhibits and injures the respiratory centre in the spinal cord, but because the incision interrupts the connection between the true respiratory centre in the fourth ventricle and the centrifugal conducting tracts to the ganglia of the nerves of the muscles of respiration.

To show more plainly that the observations made by Wertheimer on animals artificially ventilated for a considerable time do not go to prove in the least the existence of spinal respiratory centres, I allowed the cat (which for 10 days had not breathed with the right side) after the second operation to breathe for three hours by means of artificial respiration.   Then, as in the case of the dogs, the whole body showed movement, and this was true of the side first operated on as well as that on which the operation had just been performed.   Limbs and vertebræ bent and straightened, and the muscles of the neck pulled the head backwards and forwards.   The thorax moved inwards and sank without any irregularity being perceptible on either side.   The abdominal muscles contracted, and the diaphragm was the only part in which active movement could neither be seen nor felt.   Extremely strong dyspnœic movements of the head were observed.   Does Wertheimer really think that after artificial respiration has been kept up for three hours, after the cervical cord has been completely divided, that the respiratory centres have come alive again—centres which, after division on one side only and with the circulation and temperature normal, could not be brought to life again in ten days?

Further proofs against the existence of spinal-cord respiration seem to be quite unnecessary; and I could close this subject now and for ever if Langendorff, in his experiments "Regarding the effects of division of the spinal cord on one side," had not again arrived at results quite opposite to those observed by Schiff, Vulpian, and myself. Langendorff maintains that in his rabbits—he gives three examples—

notwithstanding division of the cervical cord on one side, after a few hours, respiration reappeared on the injured side. That they were few in number, but undoubtedly respiratory, is all that we are told of the character of these movements. In these rabbits in which the spinal cord had been partially divided, Langendorff also cut the phrenic nerve on the side in which respiration was normal, in order, as he thinks, to remove all doubt about the active nature of the movements of the diaphragm on the divided side. Because in animals, says he, which have been operated upon on one side the passive movement on the injured side is often so strong that one has doubts whether the movement is not taking place on both sides; on the other hand, the movement of the lowest border of the ribs is often no guide, and he would scarcely even rely on the exposed diaphragm, because, even after much practice, it is always difficult to come to a definite conclusion. For such experiments certainly rabbits are not suitable, as they respire normally entirely by the diaphragm. If the diaphragm is injured on one side this latter is moved passively by the activity of the other side; and one cannot conclude from the unequal movement, as in dogs and cats, that only one side of the thorax is moved actively. But if the whole diaphragm be injured, there appear in rabbits strong dyspnœic respiratory movements of the thorax, whereby the diaphragm is passively moved. This passive movement takes place in the whole diaphragm even when the thorax only expands on one side; if this be the case, the injured side of the thorax is moved in the other direction: when the one side expands then the other contracts, and *vice versa*. These passive movements of the injured side of the thorax are also communicated to the diaphragm, so that a movement of the diaphragm takes place quite opposed to the previous movement. From this there arises such a complication in the movements that it may become impossible to make out any distinction as to which movement is active or passive. Langendorff was aware of this, because he was forced to partially resect the thorax wall itself anteriorly and posteriorly, as well as the phrenic nerve, on the side breathing normally. Even then, according to Langendorff, there remained a few indubitable diaphragm movements. But even after such a partial resection, is one quite warranted in presuming that we have still to do with passive movement, although the *medulla spinalis* was completely divided on one side? The really important test in favour of the continuation of respiration after division of the cord on one side has not been given by Langendorff, namely, to divide the other half of the cervical cord. As, according to his own decision, it only requires practice to accomplish the division so that the stoppage of respiration is confined to the one side, then surely he would have been able at least to get *one* successful case and to demonstrate it. Besides, Wertheimer's experiences speak in favour of this view. Whether and in what manner the few respirations after *sectio bulbi* observed by Langendorff are to be brought into unison with the spinal-cord respiration described by Wertheimer as so frequent, superficial, and irregular, we are not informed. Therefore so long as doubts with regard to the method pursued by Langendorff are not settled, so long as important differences between Wertheimer and Langendorff as to spinal respiration are not explained, the results obtained by Schiff, Vulpian, and myself must receive their full value. In the spinal cord, therefore, there are no *special centres for the liberation of respiration*.   ·

There is also no respiratory centre in the corpora quadrigemina or optic thalami. Martin and Booker as well as Christiani are in error when they assume that centres are situated in the second and third ventricles of the brain. No one will deny that from the higher sensory nerves, or from the *trigeminus*, reflexes can be liberated which are partly inspiratory and partly expiratory; but a transverse division of the *medulla oblongata* at the level of the *tubercula acustica* proves that the centres of respiration must be below the incision, because respiration after this division remains quite normal and acts automatically, and the will, sensory impressions, and pain have now no influence on it. The manner in which head dyspnœa appears or ceases, after division of the *medulla oblongata* at different levels, is positive proof of the absence of respiratory centres in regions situated higher up.

Further, as one can *liberate normal inspiratory movements by means of direct stimulation of the medulla oblongata as well as of the centripetal vagi in the neck with rhythmic intermittent shocks, even after the medulla oblongata had been separated from all centripetal connections, and even when respiration had previously stopped and the heart had ceased to beat*—for instance, after bleeding from the basilar artery—we must conclude that the situation of the respiratory centre lies necessarily in the *medulla oblongata*. When Langendorff asserts that because my experiments have proved that the respiratory centre situated in the fourth ventricle can only be stimulated by reflex action, and when isolated it can only liberate respiratory spasms, this speaks against the existence of a centre and in favour of conducting tracts, then this only proves that he has not read my work so carefully as I have studied the works of the worthy author.

Certainly the respiratory centre in the bulb, like every system of ganglion cells (as Kronecker and Stanley Hall discovered), has the peculiarity that, when put in action, it does not return the received stimulation in an unaltered form. Stimulation of the centre by means of rhythmic intermittent currents liberates normal respiratory movements; while stimulation of the *medulla spinalis*, no matter in what way this may be accomplished, always produces only respiratory spasms.

Lastly, Langendorff disputes the fact which I have investigated very carefully and proved by many examples—that the respiratory centre, which is automatically active, does not liberate regular rhythmic respirations, but only respiratory spasms. He is of opinion that "the rudimentary character of the remaining respirations is to be explained by the impairment caused by the necessary operations; and if one were successful, to avoid stimulation and concussion of the nerves and central parts, to avoid loss of blood, or the great fall in the tension of blood pressure, and to avoid injury to the strength of the heart, then respiration, notwithstanding the impairment, might continue in a much better condition than is really the case." It appears to me rather daring to regard the strong respiratory spasms which remain after separation of the respiratory centre as rudimentary respiratory movements. I think I have made out these spasms to be the consequence of cessation of action of the centripetal tracts. If you divide the *medulla oblongata* above the respiratory centre, then respiration remains normal. If the vagi now cease to act, it is quite the same if they be divided with a pair of scissors or with a knife, whether they be ligatured with a thread or if by means of Gad's method of the application of cold they are made suddenly devoid of sensibility and unable to conduct. In all such cases the respiratory spasms described make their appearance. On the other hand, if the vagi are first extirpated, then in animals upon which the operation of tracheotomy has been performed, and which at first breathe deeply and slowly, respiration in a short time becomes almost normal. If you now divide the *medulla oblongata* above the respiratory centre, the respiratory spasms immediately make their appearance in the same way as when first the medulla and then the vagi ceased to act. If in a healthy animal the *medulla spinalis* below the point of escape of the phrenic nerves, then the laryngeal nerves, and then the glosso-pharyngeals, be cut, respiration remains unchanged, even when the *medulla oblongata* above the respiratory centre has been divided. The same takes place when first the *medulla oblongata* and then the dorsal spinal cord is divided along with the nerves mentioned. If you then divide the vagi, respiratory spasms immediately occur. If first the vagi and then the *medulla spinalis* are divided—or, on the other hand, first the *medulla spinalis* and then the vagi—respiration remains as above described, after the vagus ceases to act, only it is more like normal respiration. Finally, if the *medulla oblongata* above the respiratory centre is now separated, respiratory spasms immediately appear.

What do these experiments prove? Nothing else than that the isolated respiratory centre is only capable of liberating respiratory spasms, and that normally the vagi, which possess tonus, convert these respiratory spasms into regular rhythmic respiratory movements. The upper tracts of the brain can take up the action of the vagi for a time, and they do this when the vagi cease to act; while the laryngei, the glossopharyngei, and the whole of the lower tracts of the spinal cord, have no tonus and exercise no influence on the respiratory centre. As already mentioned, if, during the respiratory spasms, one stimulates either the *medulla oblongata* itself or the centripetal vagi in the neck, these respiratory spasms are changed into regular respir-atory movements. The respiratory spasms often continue for hours unchanged until the respiratory centre is fatigued, 1–2–3 in the minute, without any respiratory pause, especially when now and again artificial ventilation is for a short time introduced. But when the spinal cord is simultaneously divided (which, as already mentioned, has no direct influence on respiration), it is then necessary to use artificial respiration frequently; without this, the rabbits generally die in half an hour, probably in consequence of the great reduction in the blood pressure.

When Langendorff brings as his only proof against the results of these experiments that the isolated *medulla oblongata* of the frog behaves in an opposite manner, and is capable of liberating normal respiratory movements, then I can scarcely imagine such a proof being put seriously, especially when Langendorff adds, " It is improbable that in mammals affairs would be so essentially different in arrangement." I have no cause to investigate the condition of affairs in the frog; but it is Langendorff's duty, seeing he doubts my results, to repeat the experiments on rabbits, and then give his opinion.

Langendorff himself, in company with Dr. Joseph, has confirmed my statement that after division of the medulla above the respiratory centre and of both vagi in the neck, respiratory spasms appear, and it would have been easy to ascertain that complete separation of the respiratory centre in the fourth ventricle has the same effect, and is founded purely on the cessation of action of the centripetal tracts.

Loewy also found, "After separation of the medullary centre, a remarkable change took place in the respiration, characterized by the facts that respirations were not nearly so frequent, *generally two to four respirations per minute*, that the rhythm was quite different from the normal in so far as *the inspiration was much longer than the expiration*, that the individual respirations increased in size, and that the respiration was deepened." These are nothing else than respiratory spasms! If Loewy means that these respiratory spasms were always rhythmic, and if he understands thereby that they were always of equal length, and separated by pauses of equal duration, then this must be a great mistake. If the separation of the *medulla oblongata* has left the origins of the vagi intact, then, as long as the centres are not fatigued, there are no respiratory pauses at all, but only respiratory spasms equal in depth but differing in duration. When respiratory pauses appear, the section of the *medulla oblongata* has either injured the origin of the vagus directly or indirectly, or the centres have become fatigued. If the respiratory pauses obtain the preponderance, as they do towards the end of life, not infrequently the respiratory spasms are noticed to be much shorter, equal in duration and depth, but in that case the respiratory pauses vary in length. Sometimes two respiratory spasms are observed immediately following upon one another, among respiratory pauses of varying duration. I have in my possession numerous graphic tracings which illustrate these simple variations. They are the expressions of a more or less injured respiratory centre.

In order to get correct knowledge, separation of the respiratory centre must be repeated very often, and care should be taken that the *medulla* is divided at different levels. One will then understand why Loewy sometimes only obtained short contractions of the diaphragm (but they were surely always spasmodic), which were followed by long relaxation. Further, I always thought that Langendorff would have been the last to apply to warm-blooded animals facts drawn from experiments on frogs. Max E. G. Schrader has shown conclusively that in the frog the respiratory centre lies in the *medulla oblongata*. When Schrader separated the *medulla* from the spinal cord at the level of the apex of the *calamus scriptorius* by means of an even transverse incision, then respiration of the nose, larynx, and floor of the mouth soon returned, but the muscles of the trunk also took part in the respiratory movements. When he suppressed the respiration of the anterior part of the animal, then immediately the respiration of the cord also ceased. Probably this is due to some reflex action. *When Schrader removed the whole medulla to the apex of the calamus scriptorius, then in winter the animals might be watched for weeks, and respiratory movements never made their appearance.*

I would just mention in passing that at the meeting of naturalists held this year, Steiner in his discourse mentioned that most probably in fishes the respiratory centre is situated in the *medulla oblongata*, an observation which, as Steiner remarked, had already been made by Flourens.

I think I may conclude this investigation with the same propositions that I brought forward in my work on "The Innervation of Respiration in the Rabbit." They are as follows:—

1. In the cervical cord only the central tracts of respiration run; special centres for the liberation of respiration do not exist there.

2. The centres of respiration are situated in the *medulla oblongata*, and are in close connection with the origins of the vagi.

3. The respiratory centres in the *medulla oblongata* are automatically active, as well as excitable by reflex action.

4. The automatically active centre can only liberate respiratory spasms, but no regular rhythmic respiratory movements.

LIST OF THE WORKS

INNERVATION OF RESPIRATION AND KINDRED SUBJECTS, QUOTED IN THIS TREATISE.

Aducco V, Espirazione attiva ed inspirazione passiva. Atti della R. Academia delle Scienze di Torino. Vol. xxii. 20 Marzo 1887.

v. Anrep und N. Cybulski, Ein Beitrag zur Physiologie der Nn. phrenici (Physiol-Labor. v. *Tarchanoff.* St. Petersburg). *Pflüger's* Archiv für die gesammte Phy. siologie, Bd. 33, S. 243–249. 1883.

— — Physiol. Untersuchungen im Gebiete der Athmung und der vasom. Nerven. Petersburg 1884. Russisch. *Hofmann* und *Schwalbe's* Jahresbericht pro 1884, physiol. Th. Bd. 5, S. 66.

Arloing S, Application de la méthode graphique à l'étude du mécanisme de la déglu-tition. Comptes rendus des sc. 2 Nov. 1874.

— 2me nôte, ibidem. 24 Mai 1875.

— Application de la méthode graphique à l'étude du mécanisme de la déglutition chez les mammifères et les oiseaux. Thèse etc. juillet 1887. Paris. (G. Masson.)

— Déglutition. Dictionnaire encyclopédique des sciences médicales de Dechambre pag. 234–268. Paris.

Arloing et Tripier, Contribution à la physiologie des nerfs vagues. Archiv. de physiol. norm. et pathol. pag. 732 ff. 1871–72.

— — Contribution à la physiologie des nerfs vagues (suite et fin). Archives de physiol. norm. et patholog. Mars, p. 158. 1873.

Arnemann, Versuche über Regeneration an lebenden Thieren. Göttingen 1787.

Arnold F., Lehrbuch der Physiologie. Bb. 1, Th. II, S. 219. Zürich 1836.

Arnsberger L, Bemerkungen über das Wesen, die Ursache, die pathologisch-anatomische Natur der Lungenveränderung nach Durchschneidung der beiden Lungenmagennerven am Halse. *Virchow's* Archiv, Bd. 9. 1856.

Aubert und v. Tschischwitz, Nervis vagis irritatis diaphragma num in inspiratione an in exspiratione sistitur? Dissert. inaug. Vratislaviae 1857.

— — Molesh. Unters. 1857. Bd. 3, S. 272.

Bartels, Die Respiration als vom Gehirne abhängige Bewegung und als rhythmischer Process. Breslau 1814.

Beau et Maissiat, Recherches sur le mécanisme de la respiration. Archives générales de médec. 5 série, t. 1, p. 224. 1843.

Bell Ch., Physiol. und pathol. Untersuchungen des Nervensystems. S. 114, übers. von *Romberg.* 1832.

Bennett Dowler, M.D., New Orleans. Experimental Researches, etc. 1845 and 1851.

Bernard Claude, Leçons faites au Collége de France. Union médicale nr. 75 p. 279, nr. 78 p. 309, nr. 88 p. 349. 1853.

— Leçons sur la physiologie et pathologie du système nerveux. Vol. 2, p. 382, ff. V. C. Paris 1858.

Bernheim, Du phénomène respiratoire de *Cheyne-Stokes.* Gazette hebdomadaire p. 31. 1873.

Berns, Over den invloed van verschillende gassorten op de ademhaling. Acad. Proofschrift. Leiden 1869.

Bernstein J., Ueber die Einwirkung der Kohlensäure des Blutes auf das Athmungscentrum. *Du Bois-Reymond's* Archiv. f. Physiologie, S. 313–328. 1882.

Bert Paul, Leçons sur la physiologie comparée de la respiration. Paris 1870.

— De la contractilité des poumons. Des rapports du nerf pneumogastrique avec la respiration. D'une cause non encore signalée de mort subite. Comptes rendus. Vol. 69, nr. 8. 1869.

Bichat, Recherches physiologiques sur la vie et la mort. p. 326. 2 éd. Paris, 1802.

Bidder, Beiträge zur Kenntniss der Wirkungen des Nervus laryngeus superior. *Du Bois-Reymond's* Archiv f. Physiologie, S. 492. 1865. pag. 492–507.

Biedermann W., Ueber rhythmische, durch chem. Reizung bedingte Contraction quergestreifter Muskeln. Sitzungsber. d. Wiener Akademie d. Wissensch. Bd. 82. Nov. 1880.

— Ueber rhythmische Contractionen quer gestreifter Muskeln unter dem Einflusse des constanten Stromes. (Aus dem physiol. Inst. zu Prag.) Sitzungsber. d. Wiener Akademie d. Wissensch. Abth. III. Bd. 87. März 1883.

Bieletzky N., Zur Frage über die Ursache der Apnoe. Arbeiten der naturforsch. Gesellschaft in Charkow. Bd. 14, S. 215. (Russisch.) 1881. Biologisches Centralblatt 1881.

Biot C., Contribution à l'étude du phénomène respiratoire de *Cheyne-Stokes.* Tome XXIII. Lyon médical 1876.

— Étude clinique et expérimental sur la respiration de *Cheyne-Stokes.* Lyon 1878.

Blainville, Nouveau Bullet. de la société physiol. 1808.

Blaise et Brousse, Contributions à l'étude clinique du phénomène respiratoire de *Cheyne-Stokes.* Montpellier médical p. 287. Avril 1880.

Blumberg J., Untersuchungen über die Hemmungsfunction des N. laryng. sup. dissert. inaug. Dorpat 1865.

Boas J. H., Ueber intermittirendes Athmen. Deutsches Archiv f. klinische Medicin. 1874.

Boerhavius, p. 619. Littré Mémoires de l'Académie. p. 7. 1713.

Bongers P., Beobachtungen über die Athmung des Igels während des Winterschlafes. (Aus dem physiol. Laborat. zu Königsberg. *O. Langendorff.*) *Du Bois-Reymonds* Archiv f. Physiologie, p. 325. 1884.

Bordoni L., Sul tipo respiratorio di *Cheyne-Stokes.* Dissertation. Siena 1886.

— Respirazione periodica nell'avvelenamento per Scillainae Gelsemina. Siena 1886.

— Sull' Apnea Sperimentale. Laborator. di Fisiol. di Firenze. Lo Sperimentale. Febbraio 1888.

Bötticher W., Ueber Reflexhemmung. Samml. physiol. Abhandlgn. Bd. 4, Nr. 3, Jena 1878.

Brachet, Praktische Untersuchungen über die Verrichtung des Gangliennervensystems. Uebers. v. Flies. Quedlinburg 1836.

Brandt George-Henry, Des phénomènes de contraction musculaire observée chez des individus qui ont succombé à la suite du choléra ou de la fièvre jaune. Paris 1855.

Broadbent W. H., On *Cheyne-Stokes* respiration in cerebral haemorrhage. The Lancet, March, p. 307. 1877.

Brown-Séquard, Comptes rendus. Vol. 89, p. 657–660.

— Rhythmische Bewegungen des Zwerchfells nach Durchschneidung der Phrenici.

— Comptes rendus de l'Acad. T. XXIV, p. 363. 1847.

— Recherches expérimentales sur les resultats de l'ablation des centres nerveux, et particulièrement de la moëlle allongée dans les cinq classes des vertébrés. Comptes rendus. Vol. 26, pag. 413. 1848.

— Bulletin de la société philom. p. 117. 1849.

— De la survie des batraciens et des tortues après l'ablation de la moëlle allongée. Comptes rendus de la société de Biologie. pag. 73. 1851.

— Recherches expérimentales et observations cliniques sur le rôle de l'encéphale et particulièrement de la protubérance annulaire dans la respiration. Recherches exposées dans la thèse inaugurale de J. B. Coste. Août 1851. Paris.

— Experimental researches applied to physiology and pathology. New-York 1853.

— Experimental researches on the spinal cord. Richmond 1855.

— Recherches sur les causes de mort après l'ablation de la partie de moëlle allongée qu'on nomme le point vital. Journal de Physiologie de *Brown-Séquard.* T. I, p. 217–223. 1858.

— Du rhythme dans le diaphragme et dans les muscles de la vie animale après leur séparation des ventres nerveux. Journal de la physiologie de l'homme et des animaux. Vol. 2, p. 115. Paris 1859.

— Recherches expérimentales sur la physiologie de la moëlle allongée. Ibid. vol. 3, pag. 151. Paris 1860.

— Course of lectures on the physiology and pathology of the central nervous system delivered at the Royal College of Surgeons of England. May 1858, in octav. 276 pages. Lect. 17. Philadelphia 1860.

— Faits démontrant que le cordon latéral de la moëlle épinière ne sert pas à la respiration. Compt. rendus de la société de Biologie. pag. 64, 1869. pag. 18, 1872.

— Sur l'augmentation d'énergie des mouvements respiratoires, après la section d'une moitié latérale de la moëlle épinière. Archives de physiologie normale et pathologique. Vol. 2, pag. 299. Paris 1869.

— Faits démontrant qu'il existe trois espèces de syncope etc. Archives de physiologie norm. et pathol. Vol. 2, pag. 767. Paris 1869.

— Similarité des effets produits par la section d'une moitié latérale de la moëlle épinière et par une irritation des nerfs dorsaux, sur les mouvements volontaires et sur la respiration. Comptes rendus de la société de Biologie. pag. 140. Paris 1870.

— Arrêt de la respiration par action reflexe. Comptes rendus de la société de Biologie. pag. 134, 138, 156. Paris 1871.

— Preuves que c'est par une irritation de fibres centripètes venant des racines du nerf spinal, que l'insufflation pulmonaire arrête la respiration. Comptes rendus de la société de Biologie. pag. 22. 1872.

— Recherches expérimentales et cliniques sur l'arrêt soudain de la respiration, etc. Archives of scientific and practical medicine. New York. pag. 87. 1873.

— Prolongation extraordinaire des principaux actes de la vie après cessation de la respiration. Archives de physiologie norm. et pathol. pag. 83. 1879.

L

Brunton T. Lauder. Text-book of Pharmacology, Therapeutics and Materia Medica. p. 209. Leiden 1885.

Budge J., Die Lehre vom Erbrechen. Bonn 1840.

— Untersuchungen über das Nervensystem. 2. Heft. Frankfurt a. M. 1842.

— Mémoire sur la cessation des mouvements inspiratoires provoqués par l'irritation du nerf pneumogastrique. Comptes rendus 1854. Vol. 39, p. 749 ff.

— Observationes de vi quam nervi et Phrenicus et Vagus in respirationem habeant. (Promotionsrede.) Bonn 1855.

— Ueber den Einfluss des N. vagus auf das Athemholen. *Virchow's* Archiv 1859. Bd. 16, S. 433.

— Ueber die Zwecke des Athmens. (Ein populärer Vortrag.) Weimar 1860.

— Neuere Untersuchungen über den Einfluss des N. vagus auf die Athembewegungen. *Henle* und *Pfeufer's* Zeitschr. 3. Reihe. Bd. 21. 1864.

Burdach C. F., Vom Baue und Leben des Gehirns. Leipzig 1819.

— Die Physiologie als Erfahrungswissenschaft. 6. Bd. S. 405–497. Leipzig 1840.

Burdon-Sanderson, Handbook for the physiological laboratory. p. 318. London 1873.

Burkart R., Ueber den Einfluss des N. vagus auf die Athembewegungen. *Pflüger's* Archiv für die gesammte Physiologie. Bd. I. 1868.

— Studien über die automatische Thätigkeit des Athemcentrums und über die Beziehungen desselben zum N. vagus und andern Athemnerven. *Pflüger's* Archiv für die gesammte Physiologie. Bd. XVI. S. 427. 1878.

Calmeil, Recherches sur la structure, les fonctions et le ramolissement de la moëlle épinière. Journal des Progrès. Vol. 11, p. 116. 1828.

Cheyne, Dublin Hospital Reports. Vol. 2, p. 217. 1816.

Christiani A., Ein Athmungscentrum am Boden des dritten Ventrikels. Centralblatt f. med. Wissensch. Nro 15, pag. 273. 1880.

— Ueber Athmungsnerven und Athmungscentren. Verhandl. der Berliner physiol. Gesellschaft, 21 Mai. *Du Bois-Reymond's* Archiv f. Physiologie. S. 295. 1880.

— Experimentelle Beiträge zur Physiologie des Kaninchenhirns und seiner Nerven. Monatsbericht der Berl. Akademie d. Wissensch. Februar 1881.

— Zur Physiologie des Gehirns. Verhandl. der physiol. Gesell. zu Berlin, Nro. 15, 16. 1883–84.

— Zur Kenntniss der Functionen des Grosshirns bei Kaninchen. Sitzungsber. d. Berliner Akademie d. Wissensch. Bd. 28. 1884.

— Ueber die Erregbarkeit des Athmungscentrums. Verhandlungen der physiol. Gesellschaft zu Berlin 30. Oct. 1885. *Du Bois-Reymond's* Archiv f. Physiologie. S. 181. 1886.

Chvostek Fr., Ein Fall von *Cheyne-Stokes'*scher Respiration. Wiener medicin. Wochenschrift. 1873.

Cohnheim J., Vorlesungen über allgemeine Pathologie. Bd. 2, S. 230 ff. Berlin 1882.

Columbus, De re anatomica. Lib. V. cap. XX. p. 257. Frankfurt 1593.

Coste J. B., Recherches expérimentales et observations cliniques sur le rôle de l'encéphale et particulièrement de la protubérance annulaire dans la respiration (vide *Brown-Séquard*). Thèse inaugurale, 1 Août 1851. Paris.

Cruikshank W., Versuche über die Nerven, besonders über ihre Wiedererzeugung und über das Rückenmark lebendiger Thiere. Archiv f. die Physiol. von *Reil*. Bd. 2, Heft 1. Halle 1796.

Danilewsky, Experimentelle Beiträge zur Physiologie des Gehirns. *Pflüger's* Archiv f. die gesammte Physiologie. Bd. 11, S. 128. 1875.

v. Deen, Traités et découvertes sur la physiologie de la moëlle épinière. Leyde 1841.

Delsaux E., Sur la respiration des chauves-souris pendant leur sommeil hibernal. Bull. de l'Acad. roy. de Belg. Nro. 7. 1884.

Dohmen W., Untersuchungen über den Einfluss der Blutgase, d. i. den Sauerstoff und Kohlensäure auf die Athembewegungen ausüben. *Pflüger's* Untersuchungen aus dem physiol. Institut zu Bonn. S. 83. 1865.

Duchenne (de Boulogne), De l'électrisation localisée. I édit. Paris 1853.

— Recherches électrophysiologiques et pathologiques sur le diaphragme. Comptes rendus. Vol. 36, p. 383. 1853.

— Physiologie des mouvements. p. 611 ff. Paris 1867.

Ducrotay de Blainville et Breton. *Reil's* Archiv f. Physiologie. Bd. 11, Heft 2, S. 129. Halle 1812.

Eckhard, Grundzüge der Physiologie des Nervensystems. Giessen 1854.

— Ueber den Strychnintetanus während der künstlichen Respiration. Beiträge zur Anatomie und Physiologie. Bd. 10. 1883.

Ehrlich, Ueber die Methylenblaureaction der lebenden Nervensubstanz. Deutsche Medicin. Wochenschr. Nr. 4, S.-A. S. 7. 1886.

Eichhorst, Die trophischen Beziehungen der Nn. vagi zum Herzmuskel. Berlin 1879.

Ellenberger, Die Folgen der einseitigen und doppelseitigen Lähmungen des N. vagus bei Wiederkäuern. Archiv f. wissensch. Thierheilkunde. Bd. 9, S. 128 bis 147. 1882.

Emmert A. G. F., Ueber den Einfluss der herumschweifenden Nerven auf das Athmen. *Reil's* Archiv f. Physiologie. Bd. 9, Heft 8, S. 380. 1809.

— Nachtrag zu den Beobachtungen über den Einfluss des Stimmnervens auf die Respiration, nebst einigen Bemerkungen über den sympathischen Nerven bei Säugethieren und Vögeln. *Reil's* Archiv. Bd. 11, S. 117. 1812.

Eulenkamp T. H. P., De diaphragmate. Bonn 1856.

Ewald A., Zur Kenntniss der Apnoë. Archiv. f. die gesammte Physiologie. Bd. 7, S. 575. 1873.

Fabricius Hieronymus, De respiratione. Palav. 1615. S. 4.

Falk F., Ueber Beziehungen der Hautnerven zur Athmung. *Du Bois-Reymond's* Archiv f. Physiologie. S 455. 1884.

— Uber den mechanismus der Schluckbewegung (mit H. Kronecker) Verhandl. der physiol. Gesellschaft zu Berlin (30 April). Nr. 13. 1879–80. *Du Bois-Reymond's* Archiv f. Physiologie, 1880.

Fano G., Recherches expérimentales sur un nouveau centre automatique dans le tractus bulbospinal. Archives italiennes de Biologie. T. III. p. 365. 1883.

— Sulla respirazione periodica e sulle cause del ritmo respiratorio. . Lo sperimentale VI e VII. (giugno e luglio) 1883.

— Ancora sulla respirazione periodica e sulle cause del ritmo respiratorio. Lo sperimentale. Febbraio 1884.

— Sui movimenti respiratori del Champsa Lucius. Lo Sperimentale. Marzo 1884.

— Sulla natura funzionale del Centro respiratorio et sulla respirazione periodica, Lo sperimentale. Gennaio 1886.

Ferrier D., Die Functionen des Gehirns. Uebers. v. *Obersteiner.* Braunschweig 1879, S. 29 ff.

Fick A., Einige Bemerkungen über den Mechanismus der Athmung. Festschrift des Vereins für Naturkunde zu Cassel. 1886.

Filehne W., Ein Beitrag zur Physiologie der Athmung und Vasomotion. *Du Bois-Reymond's* Archiv f. Physiologie. S. 235–244. 1879.

— Ueber die Einwirkung des Morphins auf die Athmung. (Aus dem physiol. Institut zu Erlangen.) Leipzig 1879. Archiv für exper. Path. etc. Bd. x. pag. 442; Bd. xi. pag. 45.

— Das *Cheyne-Stokes'*sche Athmen. Zeitschr. f. klin. Medic. Bd. 2, Heft 2. 1880.

Flint A., On the cause of the movements of ordinary respiration. Are these movements reflex? Brain. p. 43. 1881.

Flourens P., Recherches expérimentales sur les propriétés et les fonctions du système nerveux dans les animaux vertébrés. Paris 1842.

— Note sur le point vital de la moëlle allongée. Comptes rendus des séances de l'Académie des sciences de Paris. Octobre. p. 437. Gaz. médicale de Paris nr. 45, p. 694. 1851.

— Nouveaux détails sur le nœud vital. Comptes rendus. 22 Nov. 1858.

— Déterminations du nœud vital ou point prémier moteur du mécanisme respiratoire dans les vertébrés à sang froid. Comptes rendus T. 54, p. 314. 1862.

Fodera, Recherches expérimentales sur le système nerveux. Journal de physiol. expériment. Vol. 3, p. 198. 1823.

Fowelin C., De causa mortis post nervos vagos dissectos. (Unter *Bidder* und *Buchheim.*) Dorpati 1851.

François-Frank, Travaux du laboratoire de Marey. T. IV, p. 281 sq. Paris 1880.

Fräntzel O., Ueber das *Cheyne-Stokes'*sche Respirationsphänomen. Berliner klin. Wochenschr. Nr. 27. 1869.

Franz C., Über künstliche Athmung. *Du Bois-Reymond's* Archiv f. Physiologie. pag. 398–415. Jhrg. 1879.

Frédericq Léon, Sur la théorie de l'innervation respiratoire. Bulletins de l'Acad. royale de Belgique II Série T. 47, No. 4. Bruxelles 1879.

— Expériences sur l'innervation respiratoire. *Du Bois-Reymond's* Archiv f. Physiologie. Supplem. p. 51–68. 1883.

Friedländer u. Herter, Ueber die Wirkung des Sauerstoffmangels auf den thierischen Organismus. S. 19–51. Zeitschr. f. physiol. Chemie Bd. 3. 1879.

Frost W. A., A case of apoplexy respiration of *Cheyne-Stokes.* The Lancet, II. p. 238. 1877.

Gad J., Ueber einen neuen Pneumatographen. *Du Bois-Reymond's* Archiv f. Physiologie. S. 181. 1879.

— Die Regulirung der normalen Athmung. *Du Bois-Reymond's* Archiv f. Physiologie. pag. 1. 1880.

— Ueber Apnoe. Würzburg 1880 und *du Bois-Reymond's* Archiv. f. Physiol. S. 28. 1880.

— Ueber die Abhängigkeit der Athemanstrengung vom N. vagus. *Du Bois-Reymond's* Archiv f. Physiol. S. 538. 1881.

— Ueber die genuine Natur reflectorischer Athembewegung. *Du Bois-Reymond's* Archiv f. Physiol. S. 566. 1881.

Gad J., Ueber Wärmedyspnoe. Sitzungsber. der Würzb. physical.-medic. Gesellschaft. S. 82. 1881.

— Ueber hämorrhagische Dyspnoe. Verhandl. d. physiol. Gesellschaft zu Berlin. Nr. 9. 1885–86.

— Ueber automatische und reflectorische Athemcentren. Verhandl. d. physiol. Gesellschaft zu Berlin. Nr. 7, 8. 1885–86.

Galen (131–200 p. Chr.), De anatomicis administrationibus. ed. *Kühn.* lib. 8, cap. 9, p. 696. Leipzig 1821.

— De causis respirationis liber. Claudii Galeni opera omnia. Tom IV ed. *Kühn.* Leipzig 1822. p. 465–469.

Gaskell W. H., The structure, distribution, and function of the Nerves which innervate the visceral and vascular system. Journal of Physiology. Vol. 7, nr. 1. 1886.

Gierke H., Die Theile der Med. oblong., deren Verletzung die Athembewegungen hemmt und das Athemcentrum. *Pflüger's* Archiv f. die gesammte Physiol. Bd. 7, S. 583. 1873.

— Zur Frage des Athmungscentrums. Centralblatt für med. Wissenschaft. Nr. 34, pag. 593–596. 1885.

Gilchrist, the British and foreign medico-surgical review. T. 22, p. 495. 1858.

Goldstein, Ueber Wärmedyspnoe. Arb. aus dem physiol. Laboratorium der Würzb. Hochschule. S. 77–90. Würzburg 1872.

Gourewitsch A., Ueber die Beziehungen des N. olfactorius zu den Athembewegungen. Dissert. inaug. Bern 1882.

Graham J. C., Ein neues specifisch-regulatorisches Nervensystem des Athmungscentrums. (Vorläufige Mittheilung.) *Pflüger's* Archiv für die gesammte Physiol. Bd. 25, S. 379–381. 1881.

Grützner, Uber die chemische Reizung von Nerven. *Pflüger's* Archiv f. Physiologie. Bd. XVII. pag. 251. 1870.

Guttmann P., Zur Lehre von den Athembewegungen. *Du Bois-Reymond's* Archiv. f. Physiologie. S. 500–525. 1875.

Haighton J., Versuche über die Reproduction der Nerven. *Reil's* Archiv Bd. 2, Heft 1. Halle 1796.

v. Haller Alberto, De respiratione experimenta anatomica. S. 4. Lausanne 1746.

— Mémoire sur plusieurs phénomènes de la respiration. S. 12. Lausanne 1758.

— Elementa physiologiae corporis humani. Tom. 3. p. 262–265. Lausanne 1761.

Hällstén K., Zur Kenntniss der sensiblen Nerven und der Reflexapparate des Rückenmarkes. *Du Bois-Reymond's* Archiv f. Physiologie. S. 167. Bd. 38. 1886.

Haro, Mémoire sur la respiration des grenouilles et des tortues. Annales des sciences naturelles. Série II. T. 18. Paris 1842.

Hegelmaier, Die Athembewegungen beim Hirndrucke. Heilbronn 1859.

Heidenhain A., Ueber *Cyon's* neue Theorie der centralen Innervation der Gefässnerven. *Pflüger's* Archiv. Bd. 4, 1871, S. 554.

Hein J., Uber die *Cheyne-Stokes* Athmungsform. Wiener med. Wochenschrift 1877.

— Ueber die Symptome und die Pathogenese des *Cheyne-Stokes'*schen Phänomens und verwandter Athmungsformen. Deutsches Archiv f. klin. Medicin. Bd. 27. 1880.

Heinemann C. U., Nonnulla de nervo vago ranarum experimenta. Berolini 1858.

— Ueber den Respirationsmechanismus der Rana esculenta und die Störung desselben nach der Durchschneidung der Nn. vagi. *Virchow's* Archiv. Bd. 22, S. 1–38. 1861.

Heinricius G., Om förändringarna i respiration och cirkulation vid genomskärning af nervi phrenici. Finska Läkaresäll-skapets Handl. Band. 30, Häft 3. 1888.

Heitler M., Ueber das *Cheyne-Stokes'*sche Respirationsphänomen. Wiener medicin. Presse, S. 649. 1874.

Helfft, Mittheilungen aus dem Gebiete der Nervenphysiologie. Untersuchungen über die Natur des Vagus. *Oppenheim'*sche Zeitschr. Februar 1850.

v. **Helmolt**, Ueber die reflector. Beziehungen des N. vagus zu den motor. Nerven der Athemmuskeln. Dissert. inaug. Giessen 1856.

Hénocque, Études expérimentales sur les fonctions du nerf phrénique. Société de Biologie. Séance du 22 juillet, 29 juillet et 5 août 1882.

Henrijean T., Sur les effets respiratoires de l'excitation du pneumogastrique (Labor. physiol. de Liège). Archives de Biologie. Vol. 3, p. 229–234. 1882.

Herhold, Sur la manière de respirer des grenouilles. Bulletin de la société philomatique. Ann. VII. T. II. pag. 42.

Hering E., Ueber den Einfluss der Athmung auf den Kreislauf. I. Mittheil. Ueber Athembewegungen des Gefässsystems. Sitzungsber. d. Wiener Akademie. Abth. 2, S. 829. 1870.

Hering Paul, Zusammensetzung der Blutgase während der Apnoe. Dissert. inaug. Dorpat 1867.

Hering und **Breuer**, Die Selbststeuerung der Athmung durch den N. vagus. 30 April. Sitzungsber. d. Wiener Akademie. Bd. 57, Abth. II. 1868.

Hermann L., Das Verhalten des kindlichen Brust Kastens bei der Geburt. *Pflüger's* Archiv für Physiologie. Bd. 30, pag 276–287. Bonn 1883.

Holmgren Frz. on **Rosenthal**, Falks försök och dass tydning. Upsala läkarefor förhandl. Bd. 18, p. 203. 1883.

Hoppe-Seyler F., Über die Ursache der Atembewegungen. Zeitschrift f. physiol. Chemie III. S. 105. 1879.

Jolyet, Les fibres du nerf pneumogastrique dont l'excitation a pour resultat de donner naissance au phénomène réflexe, qui constitue la toux, ont une tendance à s'isoler en nerf distinct séparé du tronc du nerf pneumogastrique. Compt. rend. Soc. de Biologie. S. III p. 409–411. 1877.

Joseph M., Zeitmessende Versuche über Athmungsreflexe (Physiol. Inst. Königsberg). *Du Bois-Reymond's* Archiv f. Physiologie. S. 480–487. 1883.

Kaufmann E., Ueber einige künstl. ausgelöste Erscheinungen bei *Cheyne-Stokes'*schen Atkmungsphänomen. Prager medicin. Wochenschrift 27. Aug. und 3. Sept. 1884.

Knoll, Ueber Reflexe auf die Athmung, welche bei Zufuhr einiger flüchtiger Substanzen zu den unterhalb des Kehlkopfes gelegenen Luftwegen ausgelöst werden. Sitzungsber. d. Wiener Akademie. Bd. 29. 1873.

— Über die Wirkung von Chloroform u. Aether auf Athmung u. Blutkreislauf. Einleitung. I Mittheil. Wiener Acad. Sitzungsber. III. Bd. 74. 1876. II Mittheil. Bd. 77. 1878.

Knoll, Beiträge zur Lehre von der Athmungsinnervation. 1–3. Mittheilung. Sitzungsbericht d. Wiener Akad. Abth. 3, Bd. 85, S. 282-306; Bd. 86, S. 48–65, S. 101–120. 1882.

— Ueber unregelmässiges und periodisches Athmen. Lotos 1882.

— Beiträge zur Lehre von der Athmungsinnervation. 4. Mittheil. Sitzungsber. d. Wiener Akademie. Abth. 3, Bd. 88, S. 479–512. 1883.

— Beiträge zur Lehre von der Atheminnervation 5 Mitth. Sitzungs. Ber. der Kais. Acad. der Wiss. zu Wien. Bd. 92. III. Abth. Juliheft. 1885.

— Ueber periodische Athmungs- und Blutdrucksschwankungen. Sitzungsber. d. kaiserl. Akademie d. Wissensch. Bd. 92, Abth. 3, Dez.-Heft Jahrg. 1885.

— Beiträge zur Lehre von der Athmungsinnervation VII. Mitth. kaiserl. Acad. d. Wissensch. in Wien. Bd. 95. Abth. III. 10 März 1887.

Köhler H. J., Ueber die Compensation mechanischer Respirationsstörungen und die physiol. Bedeutung der Dyspnoe. Archiv f. experimentelle Pathol. und Pharmakol. Bd. 7. 1877.

Kohts O., Experimentelle Untersuchungen über den Husten. *Virchow's* Archiv f. path. Anat. Bd. 60, S. 191. 1874.

Kohts u. Tiegel, Einfluss der Vagusdurchschneidung auf Herzschlag und Athmung. *Pflüger's* Archiv f. die gesammte Physiologie. Bd. 13, S. 84. 1876.

Kölliker und H. Muller, Versuche über den Einfluss des Vagus auf die Respiration. Würzb. Verhandl. 1854. S. 233–235.

Kratschmer, Ueber Reflexe von der Nasenschleimhaut auf Athmung und Kreislauf. Sitzungsber. d. Wiener Akademie, Juni. S. 147. 1870.

Krimer, Untersuchungen über die nächste Ursache des Hustens. Leipzig 1819.

Kronecker H., Altes u. Neues über das Athmungscentrum. Deutsche med. Wochenschrift. Nr. 36 u. 37. 1887.

Kronecker und Mc· Guire, Ueber die Speisung des Froschherzens. Verhandl. der physiol. Gesellschaft zu Berlin. Jahrg. 11, S. 56. 1877-78.

Kronecker und Marckwald, Ueber die Athembewegung des Zwerchfells. Verhandl. d. physiol. Gesellsch. zu Berlin, 25. Juli. *Du Bois-Reymond's* Archiv f. Physiol. S. 592. 1879.

Kronecker und Stirling, Ueber die sog. Anfangszuckung. *Du Bois-Reymond's* Archiv. f. Physiol. S. 394. 1878.

Kronecker u. Meltzer, Die Bedeutung des musc. mylohyoid. für den ersten Act der Schluckbewegung. Verhandl. der physiol. Gesellsch. zu Berlin. Jahrg. 1879–1880. Nro. 13. *Du Bois-Reymond's* Archiv f. Physiologie. 1880.

— — Über die Vorgänge beim Schlucken. ibidem. Nro. 18. *Du Bois-Reymond's* Archiv 1880.

— — Über den Schluckakt u. die Rolle des Cardia bei demselben. Verhandl. der physiol. Gesell. zu Berlin. Jahrg. 1880-1881. Nro. 17 u. 18. *Du Bois-Reymond's* Archiv f. Physiologie. 1881.

— — Über den Schluckmechanismus u. dessen nervöse Hemmungen. Monatsber. der Academie der Wissenschaften zu Berlin, 24 Jan. 1881.

— — On the propagation of inhibitory excitation in the med. oblongata. Proceedings of the Royal Society. Nro. 216. 1881. (Received 18 Oct. 1881.)

— — Der Schluckmechanismus, seine Erregung u. seine Hemmung. *Du Bois-Reymonds* Archiv f. Physiologie Supplement. 1883.

Kronecker u. Stanley Hall, Die willkührliche Muskelaction. *Du Bois-Reymond's* Archiv f. Physiolog. Suppl. pag. 11–47. 1879.

Laborde, Recherches expérimentales sur quelques points de la physiologie du bulbe rachidien. Gazette médec. de Paris. nr. 5. 1878.

— Sur l'indépendance fonctionelle des phénomènes mécaniques de la respiration et des mouvements du cœur à la suite d'une piqûre légère, superficielle ou du voisinage du bulbe au niveau du bec du calamus scriptorius. Compt. rend. de la soc. de Biol. pag. 181. 1883.

Laffont, Recherches sur l'innervation respiratoire; modifications des mouvements respiratoires sous l'influence de l'anesthésie. Comptes rendus. Bd. 94, S. 578 bis 581. 1883.

Lallemand, Observations pathologiques. p. 86. Paris 1818.

Lange M., Die Athmung des Frosches in ihrer Beziehung zu den Ernährungsverhältnissen der medulla oblongata. Inaugural dissertation. Königsberg 1882.

Langen, Periodische Athmung. Wiener med. Wochenschr. Nr. 40, 41. 1882.

Langendorff O., Einfluss des N. vagus und der sensiblen Nerven auf die Athmung. Mittheil. aus dem Königsberger physiol. Instit. *Du Bois-Reymond's* Archiv f. Physiol. S. 33–67. 1878.

— Ueber Selbststeuerung der Athembewegungen. Physiol. Labor. zu Königsberg. *Du Bois-Reymond's* Archiv f. Physiologie. Suppl. S. 48–53. 1879.

— Ueber das Athmungscentrum. Med. Centralblatt Nr. 51. 1879.

— Ueber ungleichzeitige Thätigkeit beider Zwerchfellhälften. Centralblatt für med. Wissensch. Nr. 51. 1879.

— Über spinale Athmungscentren. Centralbl. f. med. Wissensch. Nro. 6. pag. 97–99. 1880.

— Studien über die Innervation der Athembewegungen. I Mitth. Über die spinalen Centren der Athmung. (mit R. Nitschmann). *Du Bois-Reymond's* Archiv. f. Physiol. pag. 518. 1880.

— Studien über die Innervation der Athembewegungen. 2. Mitth. Ueber ungleichmässige Thätigkeit beider Zwerchfellhälften. Unter Mitwirkung von *R. Nitschmann* und *H. Witzack.* S. 78. *Du Bois-Reymond's* Archiv f. Physiol. 1881.

— Studien über die Innervation der Athembewegungen. 3. Mittheil. Ueber periodische Athmung bei Fröschen. Theilweise nach Versuchen von *G. Siebert.* Ebendas. S. 241. 1881.

— Studien über die Innervation der Athembewegungen. 4. Mittheil. Periodische Athmung nach Muscarin- und Digitalinvergiftung. S. 331. *Du Bois-Reymond's* Archiv f. Physiol. 1881.

— Studien über die Innervation der Athembewegungen. 5. Mittheil. Ueber Reizung des verlängerten Markes unter Mitwirkung von *F. Gürtler.* S. 519. *Du Bois-Reymond's* Archiv f. Physiol. 1881.

— Studien über die Innervation der Athembewegungen. 6. Mittheil. Das Athmungscentrum der Insekten. *Du Bois-Reymond's* Archiv. f. Physiol. S. 80–88. 1883.

— Studien über die Innervation der Athembewegungen. 7. Mittheilung. Einige neuere Untersuchungen über den Sitz des Athemcentrums. pag. 237–253.

— Studien über die Innervation der Athembewegungen. 8. Mittheilung. Die Automatie des Athemcentrums. pag. 285–289.

Langendorff O., Studien über die Innervation der Athembewegungen. 9. Mittheilung. Uber die Folgen einer halbseitigen Abtragung des Kopfmarkes. pag. 289-295. 7-9 Mittheilung: *Du Bois-Reymond's* Archiv f. Physiologie. 3 u. 4 Heft. 1887.

Laularcié, Sur un cas de paralysie spontanée du diaphragme . . . chez le cheval. Revue vétérinaire. Toulouse Mars 1886, p. 113.

Lautenbach B. F., Are there spinal respiratory centres? Philadelphia medical Times 1879.

Lawrence, Medico Chirurgical Transactions. Vol. 5, p. 166. 1813.

Legallois, Expériences sur le principe de la vie. Paris 1812.

— Oeuvres complètes avec notes de Pariset. T. 1, p. 64. Paris 1824.

Lehwess A., De diaphragmatis usu in respiratione. Berlin 1852.

Leube W., Ein Beitrag zur Frage vom *Cheyne-Stokes'*schen Respirationsphänomen. Berliner klin. Wochenschrift. April 1870.

Liebmann, Ueber die Rhythmik der Athembewegungen. Tübingen 1856.

Lindner, De nervorum vagorum in respiratione efficacitate. Dissert. inaug. Berolini 1854.

Lockenberg E., Ein Beitrag zur Lehre von den Athembewegungen. Verhandl. der Würzb. phys.-med. Gesellschaft. Bd. 4, S. 239. 1873.

Longet, Recherches expérimentales sur les agents de l'occlusion de la glotte dans la déglutition, le vomissement et la rumination. Archives générales de médecine. Paris 1841.

— Expériences relatives aux effets de l'inhalation de l'éther sulfurique sur le système nerveux de l'homme et des animaux. Archives générales de médic. Vol. 13, p. 377. 1847.

— Anatomie und Physiologie des Nervensystems der Menschen und der Wirbelthiere; übers. v. *Dr. S. A. Hein.* Bd. 1, S. 322 u. ff. Leipzig 1847.

— Traité de Physiologie. Bd. 2. Paris 1850.

Lorry, Sur les mouvements du cerveau. Académie des sciences. Mémoires des savants étrangers. Tom 3, p. 366. 1760.

Lovén Christian, Om Naturen af de voluntära Muskelkontraktionerna. Nord. med. arkiv. Bd. 13, Nr. 5. 1881.

Löwinsohn, Experimenta de nervi vagi in respirationem vi et effectu. Dissert. inaug. Dorpati Livon 1858.

Löwitt M., Ueber das *Cheyne-Stokes'*sche Respirationsphänomen. Prager med. Wochenschrift. Nr. 47. 1880.

Loewy A., Über das Athemcentrum in der Med. oblong. u. die Bedingungen seiner Thätigkeit. Ausdem thier physiol. Laborat. der landvirthschaftl. Hochschule zu Berlin, Jahrg. 1886-87. Nro. 15. 25 Juni 1887.

— Experimentelle Studien über das Athemcentrum in der Med. oblong. u. die Bedingungen seiner Thätigkeit—Über den Tonus des Lungenvagus.—Beitrag zur Kenntniss der bei Muskelthätigkeit gebildeten Athemreizl. Archiv f. dieges. Phys. Bd. XLII. pag. 245-284. 1888.

Luchs H., Symbolae ad euergias motorias systematis nervosi cognoscendas. Vratislaviae 1847.

Luciani L., Del fenomeno di *Cheyne e Stokes* in ordine alla dottrina del ritmo respiratorio (Vol. 43, p. 341-355, p. 449-466). Lo Sperimentale 1879.

Ludwig C., Lehrbuch der Physiologie des Menschen. Bd. 2. 1861.

Mader, Zur Casuistik des *Cheyne-Stokes*'schen Respirationsphänomens. Wiener medicin. Wochenschrift. 1869.

Magendie, Précis élémentaire de physiologie. Vol. 2, p. 297. Paris 1817.
— Leçons sur les fonctions etc. T. 1, p. 285, 293 sq. und Précis élémentaire de Physiologie. Paris 1825. T. 2, p. 326.
— Leçons sur les phénomènes physiques de la vie. Vol. 1, p. 215. Vol. 2, p. 223 sq. Paris 1836.

Mangili, Saggio di osservazioni per servire alla storia di mammiferi sogetto a letargo. Milano 1807.

Marchand R., Versuche über das Verhalten von Nervencentren gegen äussere Reize. *Pflüger's* Archiv f. die gesammte Physiologie. Bd. 17, S. 511–542. 1878.

Marckwald Max, Die Athembewegungen u. deren Innervation beim Kaninchen. Zeitschrift für Biologie. Bd. xxiii. Nr. Folge. Bd. v. pag. 1–135. 1886.
— Über die Ausbreitung der Erregung u. Hemmung vom Schluckcentrum auf das Athemcentrum. Zeitschrift f. Biologie. Bd. 26. N. F. Bd. VIII.

Marckwald und Kronecker, Ueber die Auslösung der Athembewegungen. Verhdl. der physiol. Gesellsch. zu Berlin. *Du Bois-Reymond's* Archiv f. Physiol. S. 100–104. 1880.

Marcuse, De suffocationis imminentis causis et curatione. Dissert. inaug. Berolini 1858.

Marshall Hall, Memoirs on the nervous system. London 1837.

Martin H. N., Normal respirator. movements of the frog etc. Journal of physiol. Vol. 1, pag. 131–170. *Johns Hopkins University* studies. Baltimore 1879.

Martin H. N., and **Mussey Hartwell,** On the respiratory function of the internal intercostal muscles. Journal of physiology II. Nro. 1. Studies from the biological Laboratory. Baltimore 1880.

Martin and Booker, *Johns Hopkins University* studies from the biological laboratory. Baltimore 1879.

Martine, Essays of society at Edimb. Vol. 1, pag. 164. 1740.

Martius Fr., Historisch-kritische und experimentelle Studien zur Physiologie des Tetanus. *Du Bois-Reymond's* Archiv f. Physiologie. 1883.
— Ueber die Wirkung blutverdünnender Transfusion bei Fröschen. Verhandl. der physiol. Gesellschaft zu Berlin. *Du Bois-Reymond's* Archiv für Physiol. Nr. 8. 1882.
— Die Erschöpfung und Ernährung des Froschherzens. (Mit einem Zusatz von *H. Kronecker*.) S. 543–56. *Du Bois-Reymond's* Archiv f. Physiol. 1882.

Mayer S., Experimenteller Beitrag zur Lehre von den Athembewegungen. Sitzungsbericht der Wiener Akademie. Bd. 69. April. Abth. 3. 1874.
— Beitrag zur Kenntniss des Athemcentrums. Prager Zeitschr. f. Heilkunde IV. pag. 187. 1883.

Meltzer S., Geschlecht und Lungenvagus (Physiol. Inst. Berlin). Centralblatt f. medicin. Wissenschaften. S. 497–498. 1882.
— Das Schluckcentrum, seine Irradiationen u. die allgemeine Bedeutung derselben. Inaugural Dissertation. 12. Aug. 1882.
— Die Irradiationen des Schluckcentrums und ihre allgemeine Bedeutung. *Du Bois-Reymond's* Archiv. S. 209-238. 1883.

Mendelssohn A., Der Mechanismus der Respiration und Circulation oder das explicirte Wesen der Lungenhyperämie. Berlin 1845.

Mendelssohn Maurice, Beitrag zur Frage nach der directen Erregbarkeit der Vorderstränge des Rückenmarkes. *Du Bois-Reymond's* Archiv f. Physiol. S. 281. 1883.

— Ueber die Irritabilität des Rückenmarkes. *Du Bois-Reymond's* Archiv f. Physiol. S. 288. 1886.

Merkel, Zur Casuistik des *Cheyne-Stokes*'schen Respirationsphänomens. Deutsches Archiv f. klin. Medicin. Bd. 8, S. 424. 26 Mai 1871.

v. Mertschinsky P., Beitrag zur Wärmedyspnoe. Inaugural-Dissertation. Würzburg 1881.

Merunowicz, Ueber die chemischen Bedingungen für Entstehung des Herzschlages. Arbeiten aus der physiol. Anstalt zu Leipzig. S. 132. 1875.

Meyse, De l'influence de la section des nerfs pneumogastriques sur la durée de la chloroformisation. Gaz. médic. de Paris. Nr. 53, p. 527. 1851.

Miescher-Rüsch, Bemerkungen zur Lehre von den Athembewegungen. *Du Bois-Reymond's* Archiv f. Physiol. Heft 5 und 6. 1885.

Mislawsky N., Zur Lehre vom Athmungscentrum. Vorläufige Mittheilung. (Aus dem physiol. Laborat. von Prof. *N. Kowalewsky* in Kasan). Centralblatt f. medic. Wissenschaft. Nr. 27. 1885.

Moleschott J., Ueber den Einfluss des Vagus auf die Häufigkeit der Athemzüge. *Moleschott's* Untersuchungen. Bd. 9, S. 59, 71. 1863.

Moleschott J. und A. **Moriggia**, Ueber ungleichsinnige Veränderungen in der Häufigkeit der Athemzüge und der Pulsfrequenz. *Moleschott's* Untersuchungen. Bd. 9, S. 162. 1863.

Mosso A., Sul polso negativo e sui rapporti della respirazione addominale e toracica nell' uomo. Archivio per le scienze mediche. 1878.

— Ueber die gegenseitigen Beziehungen der Brust- und Bauchathmung. *Du Bois-Reymond's* Archiv f. Physiol. B. 441–468. 1878.

— Sulla circolazione del sangue nel cervello dell' uomo. R. Academia dei Lincei. Vol. 5. 1879.

— La respirazione dell' uomo sulle alte montagne. Atti della R. Academia de Torino. Torino 1884.

— La paura. Milano 1884.

— La respirazione periodica e la respirazione superflua e di lusso. Reale Accad. dei Lincei. Roma 1885.

Muller Johannes, Handbuch der Physiologie des Menschen. Coblenz 1835.

Müller W., Beiträge zur Theorie der Respiration. Wiener akad. Ber. XXXIII, S. 99. Lieb. Anm. 108. S. 257. 4. Nov. 1858.

Murri A., Sulla genesi del fenomeno di *Cheyne-Stokes*. Rivista clinica Nr. 10 e 11. 1883.

— Considerazioni sul mecanismo della funzione centrale del respiro. Rivista Clinica p. 385. 1884.

— Sul origine patologica del respiro periodico. La Rivista Medica. Dec. 1885.

Nasse Fr., Untersuchunger zur Lebensnaturlehre Halle 1818.

Nasse H., Einige Versuche über die Wirkung der Durchschneidung der Nn. vagi bei Hunden bes. in Hinsicht auf den Stoffwechsel. Archiv des Vereins für gemeinsch. Arbeit. Bd. 2, Heft 3. 1855.

Nasse H., Influence de la section du nerf pneumogastrique chez les chiens. Archiv. général. de méd. S. 362–363. (Sept.) 1856.

Nasse O., Sauerstoffmangel, ein Reiz für die nervösen Centralorgane. 1870.

Neuburger Th., Disquisitiones physiologicae. Berolini 1853.

Nitschmann R., Beitrag zur Kenntniss des Athmungscentrums. *Pflüger's* Archiv. pag. 558. Bd. XXXV. 1884.

Ollivier, Traité de la moëlle épinière. Ed. 2, p. 155. Paris 1827.

v. Ott, Ueber lebenerhaltende Transfusionen mit Pferdeserum. Verhandl. der physiol. Gesellsch. zu Berlin. Jahrg. 1881–82, Nr. 11. *Du Bois-Reymond's* Archiv f. Physiologie. S. 123. 1882.

Owsjanikow, Ueber den Stillstand des Athmungsprocesses während der Exspirationsphase bei Reizung des centralen Endes vom N. vagus. *Virchow's* Archiv. Bd. 18, S. 572. 1860.

Panizza B., Observations zootomico-physiologiques sur la respiration chez les grenouilles, les salamandres et les tortues, Annal. des sciences naturelles. Série III, T. III. Zoologie. Paris 1845.

— Sul nervo frenico e sulla bolsaggine. Gaz. med. Italiana. Lombardia Nr. 8, p. 62, 64. 1865.

Panum, Ueber Vagusdurchschneidung. Med. Neuigkeit Nov. 13. März 1857.

Passerini C., Dispnea parossistica riflessa etc. Gazzetta degli Ospitali. 1885.

Pechlin, Observationes medicae. (De Apoplexia.) Hamburg 1691.

Percy, Académie des sciences. 9 Sept. 1811, p. 259.

Pflüger E., Ueber die Ursache der Athembewegungen sowie der Dyspnoe und Apnoe. *Pflüger's* Archiv für die gesammte Physiologie. Bd. 1, S. 73. 1868.

Plateau F., Recherches expérimentales sur les mouvements respiratoires des insectes. Mémoires de l'Académie royale des sciences etc. de Belgique. Tom. XLV. 219 pag. Bruxelles 1884.

Poiseuille, Recherches sur la respiration. Comptes rendus de l'Académie de Paris. T. 41, p. 1072–1076. Gazette des Hospitaux nr. 150, p. 599–600. 1855.

— Recherches sur la respiration. Gaz. hebdomad. nr. 1, p. 7. Abeille méd. nr. 1, p. 8–9. The American journal of science. April. p. 461. 1856.

v. Preuschen F., Über die Ursachen der ersten Athembewegungen. Zeitschrift f· Geburtsh. u. Gynäk. I. pag. 353–365. 1877.

Prévost J. L. (vide *Waller A. et J. L. Prévost*).

Preyer, Ueber die Ursache der ersten Athembewegung. Sep.-Abdr. aus den Sitzungsber. der Jenaischen Gesellschaft für Medicin und Naturwissensch. 6 Febr. 1880.

— Über die erste Athembewegung des Neugeborenen. Zeitschrift für Geburtshülfe u. Gynaecologie. Bd. VII. pag. 241–253. 1882.

Provenial, Journal génér. de médecine. Bd. 3, S. 7. 1810.

Rach Ed., Quomodo medulla oblongata, ut respirandi motus efficiat, incitetur. Diss. inaug. Regiomonti 1863.

Ranvier L., Leçons d'Anatomie générale (1877–1878). p. 390. Paris 1880.

Renzi P., Saggio di Fisiologia sperimentale sui centri nervosi della vita psichica nelle quattro classi degli animali vertebrati. Annali universali vol. 189, p. 419–446 (Sett.); vol. 190, p. 3–62 (Ottobr.), p. 292–348 (Nov.), p. 479 a 559 (Dec.) 1864.

Richardson B. W., M.D., on artificial respiration. Medical Times and Gazette, 1869.

Richet C., Physiologie des muscles et des nerfs. P. 284-286. Paris 1882.

— De l'influence de la chaleur sur la respiration et de la dyspnée thermique. Comptes rendus. Bd. 99, p. 279. 1884.

Riegel Fr., Die Athembewegungen. Eine physiologisch-pathologische Studie. Würzburg 1873.

Rokitansky, Untersuchungen über die Athemnervencentra. Oesterr. med. Jahrbücher. F. 1, S. 30. 1874.

Rolando, Archives générales de médecine. Paris 1823.

Romberg M. H., Lehrbuch der Nervenkrankheiten des Menschen. Aufl. 2, Bd. 3, S. 104 ff. Berlin 1851.

Roose, Anthropologische Briefe. Leipzig 1803.

Rosenbach Ottomar, Studien über den Nv. vagus. S. 109. Berlin 1877.

— Zur Physiologie des Vagus. Centralbl. f. med. Wissenschaft. Nr. 6. 1877.

— Notiz über den Einfluss der Vagusreizung auf die Athmung. Archiv f. d. gesammte Physiol. Bd. 16, S. 502. 1878.

— Zur Lehre vom *Cheyne-Stokes*'schen Athmungsphänomen. Zeitschrift für klin. Medicin. Bd. 1, Heft 3. 1880.

— *Cheyne-Stokes.* Realencyklopädie. Bd. 3. 1880.

Rosenthal J., De l'influence du nerf pneumogastrique et du nerf laryngé supérieur sur les mouvements du diaphragme. Comptes rendus. T. 52, p. 764-756. 1861.

— Ueber den Einfluss der Nn. vagi auf die Bewegungen des Zwerchfells. Amtl. Ber. der 35. Vers. d. deutschen Naturf. u. Aerzte in Königsberg i. J. 1860. S. 122-125. Königsberg 1861.

— Ueber den Einfluss des N. vagus auf die Athembewegungen. *Du Bois-Reymond's* Archiv f. Physiologie. S. 226. 1862.

— Die Athembewegungen und ihre Beziehungen zum N. vagus. Berlin 1862.

— Studien über Athembewegungen. I. Art. *Du Bois-Reymond's* Archiv f. Physiol. S. 456. 1864.

— Studien über Athembewegungen. II. Art. *Du Bois-Reymond's* Archiv f. Physiol. S. 101. 1865.

— Studien über die Athembewegungen. III. Art. *Du Bois-Reymond s* Archiv f. Physiologie. S. 423. 1870.

— Ueber die Ursache der Athembewegungen. Berliner klin. Wochenschrift. Nr. 31. 1870.

— Bemerkungen über die Thätigkeit der automatischen Nervencentren, insbesondere über die Athembewegungen. Erlangen 1875.

— Neue Studien über Athembewegungen. I. Art. Die Wirkung der elektr. Vagusreizung auf die Athembewegungen. *Du Bois-Reymond's* Archiv f. Physiologie. 1880. Suppl. Bd. pag. 34.

— Neue Studien über Athembewegungen. II. Art. Ueber die Wirkung der elektr. Reizung des N. vagus. *Du Bois-Reymond's* Archiv f. Physiol. S. 39-60. 1881.

— Neue Studien über Athembewegungen. Anhang: Ueber unipolare Nervenreizung und falsche Nervenreizung durch Nebenleitung. S. 62. *Du Bois-Reymond's* Archiv f. Physiol. 1881.

— Athembewegungen und Innervation derselben. S. 165. Handbuch d. Physiol. v. *L. Hermann.* Bd. 4, Th. 2. Leipzig 1882.

Roth, Zur Casuistik des *Cheyne-Stokes'*schen Respirationsphänomens. Deutsch. Archiv f. klin. Medicin. Bd. 10. 1872.

Runge M., Zur Frage nach der Ursache des ersten Athemzuges der Neugeborenen. Zeitschrift für Geburtshilfe und Gynaekologie. Bd. 6, Heft 2, S. 395. 1881.

Sachs Pericle, Aneurisma aortico accompagnato dal fenomeno di *Cheyne-Stokes.* Rivista clinica. Bologna febbraio 1877.

Saloz Ch., Contribution à l'étude clinique et expérimentale du phénomène respiratoire de *Cheyne-Stokes.* Dissert. inaug. Genève 1881.

Sanders Ezn., Jetz over Apnoë en Dyspnoë. Maandblad vor Naturwetensch. p. 113-115. 1870-1871.

Schiff M., Ueber den Einfluss der Vagusdurchschneidung auf das Lungengewebe. *Vierordt's* Archiv. Bd. 9 und *Moleschott's* Untersuch. Bd. 8, S. 313. 1850.

— Ueber die Functionen der hinteren Stränge des Rückenmarkes. *Moleschott's* Untersuchungen. Bd. 4, S. 84-86. 1858.

— Expériences relatives à la sensibilité tactile des cordons postérieurs de la moëlle. Gaz. méd. Août nr. 32, p. 648. 1858.

— Lehrbuch der Physiologie. S. 290 u. 323. 1858-1859.

— Le nerf laryngé est-il un nerf suspensif? Comptes rendus. T. 53, p. 285-288 et 330-333. 1861.

— Ueber die angebliche Hemmungsfunction des Nv. laryngeus sup. *Moleschott's* Untersuchungen. Bd. 8, S. 225-246. 1862.

— Influence du nerf spinal sur les mouvements du coeur. Comptes rendus. T. 58, p. 619. 1864.

— Einfluss des verlängerten Markes auf die Athmung. *Pflüger's* Archiv für die gesammte Physiologie. Pag. 624, Bd. III. 1870.

— Bericht über einige Versuchs reihen. Einfluss des verlängerten Markes auf die Athmung. S. 225. *Pflüger's* Archiv f. d. gesammte Physiologie. Bd. 4. 1871.

— Lezioni del sistema nervoso encefalico. Firenze 1873.

— Neue Versuche über die Erregbarkeit des Rückenmarkes. *Pflüger's* Archiv für die gesammte Physiologie. Bd. 38, S. 182. 1886.

Schipiloff K., Ueber den Einfluss der Nerven auf die Erweiterung der Pupille bei Fröschen. Bericht von *Schiff. Pflüger's* Archiv f. d. gesammte Physiologie. Bd. 38, (Febr. 1886), S. 258, 259.

Schmiedeberg O., Ueber die pharmakologischen Wirkungen und die therapeutische Anwendung eines Carbominsäure — Ester (Urethan). Archiv f. exper. Pathologie und Pharmakologie. S. 10. 1885.

Schrader Max E. G., Zur Physiologie des Froschgehirnes. Vorläufige Mittheilung. *Pflüger's* Archiv für Physiologie. Bd. 41, pag. 75-90. 26 Juli 1887.

Schreiber J., Ueber die Functionen des N. phrenicus. *Pflüger's* Archiv für die gesammte Physiologie. Bd. 31, S. 577-600. 1883.

Schröder van der Kolk, Bau und Physiologie der Med. spinal. u. obl. übers. von Theile. 1859.

v. Schroff, Ueber spinale Athemcentra. Wiener med. Jahrbücher. S. 324. 1875.

Schwartz H., Die vorzeitigen Athembewegungen. Leipz. 1858. Archiv f. Gynaekologie. Bd. 1, S. 361.

-- Die vorzeitigen Athembewegungen. Ein Beitrag zur Lehre von den Einwirkungen des Geburtsactes auf die Frucht. Leipzig 1858.

Schwartz H., Hirudruck und Hautreize in ihrer Wirkung auf den Fötus. Archiv für Gynaekologie. Bd. 1. 1870.

Séquard (vide *Brown-Séquard*).

Sihler Ch., On the so-called heat dyspnoea. Journal of Physiology, II. Nr. 3. Studies from the Biological laboratory. Baltimore 1880.

— Further observations on heat dyspnoea. Journal of Physiology.

Sklarek G., De respirationis frequentia dissectis nervis laryngeis. Berolini 1858.

Snellen H., Onderzoek. gedaan in het physiolog. Laborat. der Utrechtsche Hoogeskool. Jaar 7. Utrecht 1854–55.

— Einfluss des Vagus auf die Athembewegungen. Nederl. Lancet 1854 en Jan. 1885. Prager Vierteljahrschrift 1855.

Sokolow O., u. **Luchsinger** B., Zur Lehre von dem *Cheyne-Stokes*'schen Phänomen. (Physiologisches Laboratorium der Thierarzneischule in Bern.) *Pflüger's* Archiv XXIII. pag. 283. 1881.

Solaro A., Sull' origine del respiro periodico. Riforma med. Napoli. Tom. 2. p. 15, 22. 1886.

Spode O., Ueber optische Reflexhemmung. Physiol. Institut zu Königsberg. *Du Bois-Reymond's* Archiv f. Physiol. S. 113–118. 1879.

Steiner, Ueber Functionen des N. vagus. *Du Bois-Reymond's* Archiv für Physiologie. S. 577. 1878.

— Schluckcentrum und Athmungscentrum. *Du Bois-Reymond's* Archiv f. Physiologie. S. 57–79. 1883.

— Untersuchungen über die Physiologie des Froschhirnes. Braunschweig 1885.

— Die gegenseitige Vorknüpfung der Zentren des verlängerten Markes. Biologisches Centralblatt. Nro. 22, VII. Bd. pag. 678–81. 1887.

Stirling W., Ueber die Summation elektrischer Hautreize. Ber. d. Comm. sächs. Ges. d. Wissensch. S. 372. Leipzig 1874.

— On the reflex functions of the spinal cord. Edinburgh Medical Journal for April and June 1876.

Stokes, Diseases of the Heart and the Aorta. S. 324. Dublin 1854.

Stricker, Respirationsbewegungen nach Durchschneidung der Med. obl. bei mit Antiarin vergifteten Hunden. Sitzungsber. der Wiener Akademie. Bd. 75, S. 8. 1877.

Swammerdam, Tractatus physico-anatomico-medicus de respiratione usuque pulmonum etc. Lugd. Batavorum 1679.

Tanquerel des Planches, Traité des maladies de plombe. T. 2, p. 61, 130, 140. 1830.

Tawnson R., Observationes physiologicæ de amphibiis. Pars prima de respiratione. Göttingen 1794, continuatio 1795.

Thiry L., Des causes des mouvements respiratoires et de la Dyspnée. Recueuil des travaux de la société médicale allemande de Paris. p. 55, 73. 1865.

Tizzoni G., Sulle alterazioni istologiche del Bulbi e dei vaghi che detrimano il fenomeno di *Cheyne-Stokes*. Memorie dell' Acad. delle Scienze dell' Instituto di Bologna. Ser. IV. Tom. V. 1884.

— Nuovi studi sulle alterazioni del Bulbo nel fenomeno di *Cheyne-Stokes*. Ibidem. Serie IV. Tom. VIII. 1886.

Toussaint J. A., Application de la méthode graphique à la détermination du mécanisme de la déglutition dans la rumination. Compt. rend. des sciences 24

176   LIST OF THE WORKS

Août pag. 532–537. 1874. Archives de physiologie norm. et pathol. Mars-Avril 1875.

Traube L., Beiträge zur exper. Pathol. u. Physiol. Heft 1, S. 101. 1846.

— Beitrag zur Lehre von den Erstickungs- (dypsnoetischen) Erscheinungen am Respirationsapparat. Heft 2, Beiträge zur exper. Pathol. und Physiol. 1846.

— Zur Physiologie des N. vagus. Zeitung des Vereins f. Heilkunde. Nr. 5. 1847.

— Zur Physiologie der Respiration. Allgem. med. Centralztg. Nr. 38. Mai. S. 297. Nr. 39, S. 305. (Gesammelte Beiträge: I.) 1862.

— Ueber periodische Thätigkeitsanstrengungen des vasomotorischen und Hemmungsnervencentrums. Centralbl. für d. med. Wissenschaften. S. 881. 1865.

— Die Symptome der Krankheiten des Respirations- und Circulationsapparates, Vorlesungen gehalten an der Friedrich-Wilhelms-Universität zu Berlin.— Berlin 1867.

— Ueber das Cheyne-Stokes'sche Respirationsphänomen. Berliner klin. Wochenschrift Nr. 27. 1869. (Gesammelte Beiträge. Bd. 2, Thl. 2, Nr. 61, S. 882. Berlin 1871.)

— Zur Theorie des Cheyne-Stokes'schen Athmungsphänomens. Berliner klin. Wochenschrift Nr. 16 und 18. 1874. (Gesammelte Beiträge B. 3, S. 103. 1878.)

Treviranus G. R. und L. C., Vermischte Schriften anatom. und physiol. Inhalts. Bd. 1, S. 105. Bremen 1817–1821.

Tripier und E. Arloing, Contribution à la physiologie des nerfs Vagues. Archives de physiol. norm. et patholog. Nr. 4. 1872.

Valentin G., Lehrbuch der Physiologie des Menschen. Bd. 2, Abth. 2. Braunschweig 1848.

— Die Einflüsse der Vaguslähmung auf Lungen und Hautausdünstung. Frankfurt a. M. 1857.

— Archiv f. physiol. Heilkunde. 1858. S. 433 ff.

— Beiträge zur Kenntniss des Winterschlafes der Murmelthiere. Abth. 3. Moleschott's Untersuchungen. Abth. 4, 9 und 16 ff. 1857.

Valsalva Anton, Mar. Opera omnia Epistola XIII. Venetiis 1740.

Vierordt, Physiologie des Athmens. Karlsruhe 1845.

Vogt G., Die Respirationsbewegungen des Frosches in ihrer Abhängigkeit von der Medulla oblongata. Eckhard's Beiträge zur Anat. u. Physiol. Giessen 1861.

Volkmann A. W., Ueber die Bewegung des Athmens. S. 337. Müller's Archiv. 1841.

— Nervenphysiologie. Wagner's Handwörterbuch d. Physiologie. Bd. 2. 1844.

Vulpian A., Sur l'action qu'exercent les anesthésiques (éther sulfurique, chloroforme, chloral hydraté) sur le centre respiratoire et sur les ganglions cardiaques. Compt. rend. LXXXVI. No. 21. 1878.

— Maladies du Système nerveux. Leçons prof. à la Faculté de Medic. à Paris. IIème Vol. 1885.

Waller A. et J. L. Prévost: Note relative aux nerfs sensitifs qui président aux phénomènes reflexes de la déglutition. Compt. rendus 16 Août. Pag. 480. 1869.

— Etude relative aux nerfs sensitifs qui président aux phénomènes reflexes de la déglutition. Archive norm. et pathol. Bd. III. pag. 185–197. et pag. 343–354. Paris 1870.

Wassilieff, Wo wird der Schluck ausgelöst? Zeitschrift für Biologie. Bd. XXIV. N. F. Bd. VII. pag. 29–46. 1887.

Wedenskii N., Ueber die Athmung des Frosches (Rana temporaria). Physiolog. Laborat. d. Petersburger Univ. *Pflüger's* Archiv f. gesammte Physiol. Bd. 25, S. 129–150. 1881.

— Ueber den Einfluss der elektr. Vagusreizung auf die Athembewegungen bei Säugethieren (Physiol. Inst. Breslau). *Pflüger's* Archiv f. d. gesammte Physiologie. Bd. 27, S. 1–22. 1882.

Wegele Carl, Über die centrale Natur reflectorischer Athmungshemmung. Verhandl. der physiol. Gesellsch. zu Würzburg. N. F. XVII. Bd. Würzburg 1882.

Weiss, Beitrag zur Lehre von den Reflexen im Rückenmarke. Medicin. Jahrb. der Gesellschaft d. Aerzte in Wien. S. 485–494. 1878.

Wertheimer (Nancy), Rétablissement des mouvements respiratoires après la section de la moëlle cervicale. Semaine médicale 12. Mai. P. 196. 1886.

— Sur les centres respiratoires de la moëlle épinière. Compt. rend. Soc. de biol. p. 34–36. Compt. rend. de l'Acad. CII. Paris 1886.

— Recherches expérimentales sur les centres respiratoires de la moelle épinière. Journal de l'Anatomie et de la Physiologie norm. et pathol. Jhrg. 22, Nro 5, Sept.–Oct. pag. 458–507. 1886.

— Recherches expérimentales sur les centres respiratoires de la moëlle épinière. Deuxième mémoire. Journal de l'Anat. et de la Physiol. Jhrg. 23. (nov.–déc.) 1887. pag. 567–611.

Wilson Ph., an experiment. inquiry into the laws of the vital functions. Sev. edit. London 1818.

Winslow, Sur les mouvements de la respiration. Mémoires lu à l'Académie des sciences. 1753.

v. Wittich, Ueber die Beziehungen der Med. obl. zu den Athembewegungen bei Fröschen. *Virchow's* Archiv. Bd. 37, S. 322. 1866.

Wolff, De functionibus nervi vagi. Dissert. inaug. Berl. 1856.

Wundt W., Versuche über den Einfluss der Durchschneidung der Lungennerven auf die Respirationsorgane. S. 296. *Müller's* Archiv 1855.

Yppey, Physiol. Beobachtungen über die willkürliche und unwillkürliche Bewegung der Muskeln. (Aus dem Latein. übers. v. *Leune.*) Cap. III S. 174. (Einige das Athemholen betreffende Fragen.) Leipzig 1789.

Zander R., Die Folgen der Vagusdurchschneidung bei Vögeln. Vorläufige Mittheilung. Centralbl. f. med. Wissenschaften. Nro. 6, pag. 99–102. Nro. 7, pag. 113–115. 1879.

Zuntz, Zur Kenntniss des Vagus-Tonus (nach Versuchen des H. Dr. Loewy). Verhandl. der physiolog. Gesellsch. zu Berlin. Nro. 15–25. Juni. 1887. Jahrg. 1886–1887.

Zuntz N. und **Geppert** J., Ueber die Natur der normalen Athemreize und den Ort ihrer Wirkung. (Vorläufige Mittheilung.) *Pflüger's* Archiv f. die gesammte Physiologie. Bd. 38, S. 337. 1886.

DESCRIPTION OF PLATE.

Diaphragm respiration of a rabbit after section of the spinal cord at the level of the last cervical vertebra, the vagus nerve having been cut in the neck and the *medulla oblongata* also divided above the centre of respiration.

I. Respiration after section of the spinal cord (u); at a, extirpation of one vagus; at b, extirpation of the other vagus; c represents the respiration after division of the spinal cord and the vagi.

II. Respiratory spasms after division of the *medulla oblongata*, following the above operations, above the centre of respiration (two lines drawn through one another); at t, thoracic respirations during inspiratory spasms of the diaphragm.

www.ingramcontent.com/pod-product-compliance
Lightning Source LLC
Chambersburg PA
CBHW021711210326
41599CB00013B/1616